FLIGH

Practical Test Standards

1991

FLIGHT STANDARDS SERVICE
Washington, DC 20591

NOTE

The INTRODUCTION of this practical test book includes information that is pertinent to ***all*** flight instructor practical test standards in addition to that which applies specifically to FAA-S-8081-6A, Flight Instructor - Airplane (Single-engine and Multiengine) Practical Test Standards.

Material in FAA-S-8081-6A will be effective May 31, 1991. All previous editions of this book will be obsolete as of this date.

FOREWORD

The Flight Instructor - Airplane Practical Test Standards book has been published by the Federal Aviation Administration (FAA) to establish the standards for the flight instructor certification practical tests for the airplane category and the single-engine and multiengine classes. FAA inspectors and designated pilot examiners shall conduct practical tests in compliance with these standards. Flight instructors and applicants should find these standards helpful in practical test preparation.

David R. Harrington
Acting Director, Flight Standards Service

INTRODUCTION

The Aviation Standards National Field Office of the FAA has developed this practical test book as a standard to be used by FAA inspectors and designated pilot examiners when conducting flight instructor - airplane (single-engine) and flight instructor - airplane (multiengine) practical tests. Flight instructors are expected to use this book when preparing flight instructor applicants for practical tests.

This book can be purchased from the Superintendent of Documents, U.S. Government Printing Office, Washington, DC 20402.

The FAA gratefully acknowledges the valuable assistance provided by a nation-wide public "Job Task Analysis" team that developed the knowledge, skills, and abilities which appear in this book.

Comments regarding this book should be sent to:

U.S. Department of Transportation
Federal Aviation Administration
Aviation Standards National Field Office
Operations Support Branch, AVN-130
P.O. Box 25082
Oklahoma City, OK 73125

Practical Test Standard Concept

Federal Aviation Regulations (FAR's) specify the areas in which knowledge and skill must be demonstrated by the applicant before the issuance of a flight instructor certificate with the associated category and class ratings. The FAR's provide the flexibility that permits the FAA to publish practical test standards containing specific TASKS in which competency must be demonstrated. The FAA will revise this book whenever it is determined that changes are needed in the interest of safety. Adherence to the provisions of regulations and the practical test standards is mandatory for the evaluation of flight instructor applicants.

Flight Instructor Responsibility

An appropriately rated flight instructor is responsible for training the flight instructor applicant to acceptable standards in ***all*** subject matter areas, procedures, and maneuvers included in the TASKS within the appropriate flight instructor practical test standard. Because of the impact of their teaching activities in developing safe, proficient pilots, flight instructors should exhibit a high level of knowledge, skill, and the ability to impart that knowledge and skill to students. The flight instructor must certify that the applicant:

1. is able to make a practical application of the fundamentals of instructing;
2. is competent to teach the subject matter, procedures, and maneuvers included in the standards to students with varying backgrounds and levels of experience and ability;
3. is able to perform the procedures and maneuvers included in the standards to at least the COMMERCIAL PILOT skill level[1] (or, in the case of the Flight Instructor - Instrument applicant, to the INSTRUMENT PILOT skill level) while giving effective flight instruction; and
4. is competent to pass the required practical test for the issuance of the flight instructor certificate with the associated category and class ratings or the addition of a category and/or class rating to a flight instructor certificate.

Throughout the applicant's training, the flight instructor is responsible for emphasizing the performance of, and the ability to teach, effective visual scanning and collision avoidance procedures. These areas are covered in AC 90-48, Pilot's Role in Collision Avoidance; AC 61-21, Flight Training Handbook; AC 61-23, Pilot's Handbook of Aeronautical Knowledge; and the Airman's Information Manual.

Examiner[2] Responsibility

The examiner who conducts the practical test is responsible for determining that the applicant meets acceptable standards of knowledge, skill, and teaching ability in the selected TASKS. The examiner makes this determination by accomplishing an action that is appropriate to each selected TASK and includes an evaluation of the applicant's:

1. ability to apply the fundamentals of instructing;
2. knowledge of, and ability to teach, the subject matter, procedures, and maneuvers covered in the TASKS;

[1]The term "Commercial Pilot Skill Level" is defined, for the purpose of this publication, as performing a procedure or maneuver within the tolerances listed in the FAA Commercial Pilot Practical Test Standards. If the maneuver appears only in a Private Pilot Practical Test Standard, the term means that the applicant's performance is expected to be "more precise" than that indicated by the stated tolerances. This "more precise" performance must be determined by the examiner through the exercise of subjective judgment.

[2]The word "examiner" is used throughout this standard to denote either the FAA inspector or FAA designated pilot examiner who conducts an official flight test.

3. ability to perform the procedures and maneuvers included in the standards to at least the COMMERCIAL PILOT skill level (or in the case of the Flight Instructor - Instrument applicant, to the INSTRUMENT PILOT skill level) while giving effective flight instruction; and
4. ability to analyze and correct common errors related to the procedures and maneuvers covered in the TASKS.

It is intended that oral testing be used at any time during the practical test to determine that the applicant can instruct effectively and has a comprehensive knowledge of the TASKS and their related safety factors.

Throughout the flight portion of the practical test, the examiner will evaluate the applicant's use of visual scanning and collision avoidance procedures, and the applicant's ability to teach those procedures.

Flight Instructor Practical Test Book Description

This book contains the practical test standards for Flight Instructor - Airplane (Single-engine and Multiengine). Other flight instructor practical test books include:

FAA-S-8081-7, Flight Instructor - Rotorcraft (Helicopter and Gyroplane)

FAA-S-8081-8, Flight Instructor - Glider

FAA-S-8081-9, Flight Instructor - Instrument (Airplane and Helicopter)

The loose-leaf feature of this book allows the incorporation of changes which will be sold, as required. This will permit the dissemination of information concerning changes in regulations, pilot certification procedures, and other areas related to safety upon which emphasis should be placed.

The Flight Instructor Practical Test Standards include the AREAS OF OPERATION and TASKS for the issuance of an initial flight instructor certificate and for the addition of category and/or class ratings to that certificate.

Initial Flight Instructor Certification

An applicant who seeks initial flight instructor certification will be evaluated in all AREAS OF OPERATION of the standards appropriate to the rating(s) sought. The evaluation will include at least one TASK in each AREA OF OPERATION and will ***always*** include the required TASKS.

Addition of Aircraft Category And/Or Class Ratings To A Flight Instructor Certificate

An applicant who holds a flight instructor certificate and seeks an additional aircraft category and/or class rating will be evaluated in at least the AREAS OF OPERATION and TASKS that are unique and appropriate to the rating(s) sought (see table at the beginning of each standard). At the discretion of the examiner, the applicant's competence in ***all*** AREAS OF OPERATION may be evaluated.

CATEGORY AND/OR CLASS RATINGS SOUGHT	*APPLICABLE BOOK AND SECTION*
ASE	*FAA-S-8081-6, Section 1*
AME	*FAA-S-8081-6, Section 2*
RH	*FAA-S-8081-7, Section 1*
RG	*FAA-S-8081-7, Section 2*
G	*FAA-S-8081-8*
IA	*FAA-S-8081-9*
IH	*FAA-S-8081-9*

LEGEND

ASE	*Airplane Single-Engine*
AME	*Airplane Multiengine*
RH	*Rotorcraft Helicopter*
RG	*Rotorcraft Gyroplane*
G	*Glider*
IA	*Instrument Airplane*
IH	*Instrument Helicopter*

NOTE: When administering a test based on FAA-S-8081-6A, Sections 1 and 2, the TASKS appropriate to the class airplane (land or sea) used for the test should be included.

Flight Instructor Practical Test Standard Description

AREAS OF OPERATION are phases of the practical test. In this practical test book, the first AREA OF OPERATION is Fundamentals of Instructing; the last is Postflight Procedures. However, the examiner may conduct the practical test in any sequence that results in a complete and efficient test.

TASKS are knowledge areas, flight procedures, or maneuvers appropriate to an AREA OF OPERATION. The abbreviation(s) within parentheses immediately following a TASK refer to the category and/or class aircraft appropriate to that TASK. The meaning of each abbreviation follows:

ASEL	Airplane, Single Engine Land
AMEL	Airplane, Multiengine Land
ASES	Airplane, Single-Engine Sea
AMES	Airplane, Multiengine Sea
RH	Rotorcraft - Helicopter
RG	Rotorcraft - Gyroplane
G	Glider
IA	Instrument - Airplane
IH	Instrument - Helicopter

REFERENCE identifies the publication(s) that describes the TASK. Descriptions of TASKS and maneuver tolerances are not included in the flight instructor standards because this information can be found in references listed for each TASK. Publications other than those listed may be used as references if their content conveys substantially the same meaning as the referenced publications. References listed in the four flight instructor practical test books include the current revisions of the following publications:

FAR Part 61	Certification: Pilots and Flight Instructors
FAR Part 91	General Operating and Flight Rules
FAR Part 97	Standard Instrument Approach Procedures
NTSB Part 830	Notification and Reporting of Aircraft Accidents and Incidents
AC 00-2	Advisory Circular Checklist
AC 00-6	Aviation Weather
AC 00-45	Aviation Weather Services
AC 60-14	Aviation Instructor's Handbook
AC 61-13	Basic Helicopter Handbook
AC 61-21	Flight Training Handbook
AC 61-23	Pilot's Handbook of Aeronautical Knowledge
AC 61-27	Instrument Flying Handbook
AC 61-65	Certification: Pilots and Flight Instructors

AC 61-67	**Stall and Spin Awareness Training**
AC 61-84	**Role of Preflight Preparation**
AC 61-94	**Pilot Transition Course for Self-Launching or Powered Sailplanes (motorgliders)**
AC 67-2	Medical Handbook for Pilots
AC 90-48	Pilots' Role in Collision Avoidance
AC 91-13	Cold Weather Operation of Aircraft
AC 91-23	Pilot's Weight and Balance Handbook
FAA-S-8081-1	Private Pilot Practical Test Standards
FAA-S-8081-2	Commercial Pilot Practical Test Standards
FAA-S-8081-4	Instrument Rating Practical Test Standards
AIM	Airman's Information Manual
IAP's	Instrument Approach Procedures (charts)
SID's	Standard Instrument Departures
STAR's	Standard Terminal Arrivals
AFD	Airport Facility Directory
NOTAM's	Notices to Airmen
	Pertinent Pilot Operating Handbooks and FAA-Approved Flight Manuals

Each TASK has an Objective. The examiner determines that the applicant meets the TASK Objective through the demonstration of competency in various elements of knowledge and/or skill. The Objectives of TASKS in certain AREAS OF OPERATION, such as Fundamentals of Instructing and Technical Subject Areas, include ***only*** knowledge elements. The Objectives of TASKS in the AREAS OF OPERATION that include elements of skill as well as knowledge also include common errors which the applicant must be able to describe, recognize, analyze, and correct.

The Objective of a TASK that involves pilot skill consists of four parts. Those four parts include determination that the applicant exhibits:

1. instructional knowledge of the elements of a TASK. This is accomplished through descriptions, explanations, and simulated instruction;
2. instructional knowledge of common errors related to a TASK, including their recognition, analysis, and correction;
3. the ability to demonstrate and simultaneously explain the key elements of a TASK. The TASK demonstration must be to the COMMERCIAL PILOT skill level (or, in the case of the Flight Instructor - Instrument applicant, to the INSTRUMENT PILOT skill level); the teaching techniques and procedures should conform to those set forth in AC 60-14, Aviation Instructor's Handbook; AC 61-21, Flight Training Handbook, and AC 61-27, Instrument Flying Handbook; and
4. the ability to analyze and correct common errors related to a TASK.

Use Of The Practical Test Standards Book

All of the procedures and maneuvers in the Private Pilot, Commercial Pilot, and Instrument Rating Practical Test Standards have been included in the Flight Instructor Practical Test Standards. However, to permit the completion of the practical test for initial certification within a reasonable timeframe, the examiner will select one or more TASKS in each AREA OF OPERATION. In certain AREAS OF OPERATION, there are ***required*** TASKS which the examiner must select. These ***required*** TASKS are identified by **NOTES** immediately following the AREA OF OPERATION titles.

The term "instructional knowledge" means the "what," "why," and "how" of a subject matter topic, procedure, or maneuver. It also means that the flight instructor applicant's discussions, explanations, and descriptions should follow the recommended teaching procedures and techniques explained in AC 60-14, Aviation Instructor's Handbook.

The FAA requires that all practical tests be conducted in accordance with the appropriate Flight Instructor Practical Test Standards and the policies set forth in the INTRODUCTION. The flight instructor applicant must be prepared to demonstrate the ability to instruct effectively in ***all*** TASKS included in the AREAS OF OPERATION of the appropriate practical test standards.

In preparation for the practical test, the examiner will develop a "plan of action." The "plan of action" for an initial certification test will include one or more TASKS in each AREA OF OPERATION and will ***always*** include the required TASKS. If the applicant is unable to perform a TASK listed in the "plan of action" due to circumstances beyond his/her control, the examiner may substitute another TASK from the applicable AREA OF OPERATION.

The "plan of action" for a test administered for the addition of an aircraft category and/or class rating to a flight instructor certificate will include the required AREAS OF OPERATION as indicated in the table at the beginning of each standard. The required TASKS appropriate to the rating(s) sought must also be included. The examiner will select at least one TASK in each AREA OF OPERATION. In some instances, notes identify additional ***required*** TASKS. ***Any TASK selected will be evaluated in its entirety.***

NOTE: AREA OF OPERATION IX, Stalls, Spins, and Maneuvering During Slow Flight, contains TASKS referred to as "proficiency" and "demonstration." The intent of TASKS A and B (proficiency) is to ensure that the flight instructor applicant is tested on proficiency for the purpose of teaching these TASKS to students. The intent of TASKS C, D, and E (demonstration) is to ensure that the flight instructor applicant is knowledgeable of the maneuvers and can demonstrate them to students for both familiarization and stall/spin awareness purposes.

With the exception of the ***required*** TASKS, the examiner will not tell the applicant in advance which TASKS will be included in the "plan of action." The applicant should be well prepared in ***all*** knowledge and skill areas included in the standards. Throughout the flight portion of the practical test, the examiner will evaluate the applicant's ability to simultaneously demonstrate and explain procedures and maneuvers, and to give flight instruction to students at various stages of flight training and levels of experience.

The purpose for including common errors in certain TASKS is to assist the examiner in determining that the flight instructor applicant has the ability to recognize, analyze, and correct such errors. ***The examiner will not simulate any condition that may jeopardize safe flight or result in possible damage to the aircraft.*** The common errors listed in the TASK Objectives may or may not be found in the TASK References. However, the FAA considers their frequency of occurrence justification for their inclusion in the TASK Objectives.

The examiner will place special emphasis on the applicant's demonstrated ability to teach precise aircraft control and sound judgment in decision making. The evaluation of the applicant's ability to teach judgment will be accomplished by asking the applicant to describe the oral discussions and the presentation of practical problems that would be used in instructing students in the exercise of sound judgment. The examiner will also emphasize the evaluation of the applicant's demonstrated ability to teach stall/spin awareness, spatial disorientation, collision avoidance, checklist usage, use of distractions, and any other areas directed by future revisions of the standards.

Flight Instructor Practical Test Prerequisites

An applicant for a flight instructor initial certification practical test is required by regulation to:

1. have passed the appropriate flight instructor written test(s) since the beginning of the 24th month before the month in which he or she takes the practical test;
2. hold a commercial pilot or airline transport pilot certificate with an aircraft rating appropriate to the flight instructor rating sought;
3. hold an instrument rating if applying for an airplane or an instrument instructor rating;
4. have the prescribed aeronautical experience and instruction for a flight instructor certificate with the rating sought;
5. have reached the age of 18 years; and
6. obtain a written statement from an appropriately certificated and qualified flight instructor certifying that the applicant has been given flight instruction in the items required by FAR Section 61.187(a) in preparation for the practical test within 60

days preceding the date of application. The statement shall also state that the instructor finds the applicant competent to pass the practical test, and that the applicant has satisfactory knowledge of the subject area(s) in which a deficiency was indicated on the airman written test report.[1]

An applicant holding a flight instructor certificate who applies for an additional rating on that certificate must:

1. hold an effective pilot certificate with ratings appropriate to the flight instructor rating sought;
2. have at least 15 hours as pilot in command in the category and class aircraft appropriate to the rating sought;
3. have passed the written test prescribed for the issuance of a flight instructor certificate with the rating sought since the beginning of the 24th month before the month in which he or she takes the practical test; and
4. obtain a written statement from an appropriately certificated and qualified flight instructor certifying that the applicant has been given flight instruction in the items required by FAR Section 61.187(a) in preparation for the practical test within 60 days preceding the date of application. The statement shall also state that the instructor finds the applicant competent to pass the practical test, and that the applicant has satisfactory knowledge of the subject area(s) in which a deficiency was indicated on the airman written test report.[1]

Aircraft and Equipment Required For The Practical Test

The flight instructor applicant is required by FAR Section 61.45 to provide an airworthy, certificated aircraft for use during the practical test. This section further requires that the aircraft:

1. have fully functioning dual controls except as provided in FAR Section 61.45; and
2. be capable of performing all appropriate TASKS for the flight instructor rating sought and have no operating limitations which prohibit the performance of those TASKS. A complex airplane must be furnished for the performance of takeoff and landing maneuvers, and appropriate emergency procedures. A complex landplane is one having retractable gear, flaps, and controllable propeller. A complex seaplane is one having flaps and controllable propeller.

[1]AC 61-65, Certification: Pilots and Flight Instructors, states that the instructor may sign the recommendation on the reverse side of FAA Form 8710-1, Airman Certificate and/or Rating Application, in lieu of the previous statement, provided all appropriate FAR Part 61 requirements are substantiated by reliable records.

Satisfactory Performance

The practical test is passed if, in the judgment of the examiner, the applicant demonstrates satisfactory performance with regard to:

1. knowledge of the fundamentals of instructing;
2. knowledge of the technical subject areas;
3. knowledge of the flight instructor's responsibilities concerning the pilot certification process;
4. knowledge of the flight instructor's responsibilities concerning logbook entries and pilot certificate endorsements;
5. ability to demonstrate the procedures and maneuvers selected by the examiner to at least the COMMERCIAL PILOT skill level (or in the case of the Flight Instructor - Instrument applicant, to the INSTRUMENT PILOT skill level) while giving effective instruction;
6. competence in teaching the procedures and maneuvers selected by the examiner;
7. competence in describing, recognizing, analyzing, and correcting common errors simulated by the examiner; and
8. knowledge of the development and effective use of a course of training, a syllabus, and a lesson plan.

Unsatisfactory Performance

If, in the judgment of the examiner, the applicant does not meet the standards of performance of any TASK performed, the applicable AREA OF OPERATION is considered unsatisfactory and; therefore, the practical test is failed. The examiner or applicant may discontinue the test at any time when the failure of an AREA OF OPERATION makes the applicant ineligible for the certificate or rating sought. The test will be continued only with the consent of the applicant. If the test is discontinued, the applicant is entitled credit for only those AREAS OF OPERATION satisfactorily performed. However, during the retest and at the discretion of the examiner, any TASK may be re-evaluated, including those previously considered satisfactory. Specific reasons for disqualification are:

1. failure to perform a procedure or maneuver to the COMMERCIAL PILOT skill level (or in the case of the Flight Instructor - Instrument applicant, to the INSTRUMENT PILOT skill level) while giving effective flight instruction;
2. failure to provide an effective instructional explanation while demonstrating a procedure or maneuver (explanation during the demonstration must be clear, concise, technically accurate, and complete with no prompting from the examiner);
3. any action or lack of action by the applicant which requires corrective intervention by the examiner to maintain safe flight;
4. failure to use proper and effective visual scanning techniques to clear the area before and while performing maneuvers.

SECTION 2

FLIGHT INSTRUCTOR AIRPLANE - MULTIENGINE

Practical Test Standards

ADDITION OF A MULTIENGINE CLA
AIRPLANE CATEGORY RATING, IF AP
FLIGHT INSTRUCTOR CERTIFICATE.

REQUIRED AREAS OF OPERATION	FLIGHT INSTRUCTOR CERTIFICATE AND RATING					
	ASE	RH	RG	G	IA	Ih
I	NO	NO	NO	NO	NO	NO
II	NO	YES	YES	YES	YES	YES
III	NO	YES	YES	YES	YES	YES
IV	NO	NO	NO	NO	NO	NO
V	YES	YES	YES	YES	YES	YES
VI	NO	YES	YES	YES	YES	YES
VII	NO	YES	YES	YES	YES	YES
VIII	YES	YES	YES	YES	YES	YES
IX	NO	YES	YES	YES	YES	YES
X	YES	YES	YES	YES	YES	YES
XI	NO	YES	YES	YES	NO	YES
XII	NO	YES	YES	YES	YES	YES
XIII	NO	YES	YES	YES	YES	YES
XIV	YES	YES	YES	YES	YES	YES
XV	YES	YES	YES	YES	YES	YES
XVI	NO	YES	YES	YES	YES	YES

NOTE: If an applicant holds more than one rating on a flight instructor certificate and the table indicates both a "yes" and a "no" for a particular AREA OF OPERATION, the "no" entry applies. This is logical since the applicant has satisfactorily accomplished the AREA OF OPERATION on a previous flight instructor practical test. At the discretion of the examiner, the applicant's competence in all AREAS OF OPERATION may be evaluated.

CONTENTS

A. CHECKLISTS:

1. Applicant's Practical Test Checklist 2-vii
2. Examiner's Checklist 2-ix

B. AREAS OF OPERATION

I. FUNDAMENTALS OF INSTRUCTING

A. The Learning Process 2-1
B. The Teaching Process 2-1
C. Teaching Methods 2-2
D. Evaluation 2-2
E. Flight Instructor Characteristics
and Responsibilities 2-3
F. Human Factors 2-3
G. Planning Instructional Activity 2-4

II. TECHNICAL SUBJECT AREAS

A. Aeromedical Factors 2-5
B. Visual Scanning and Collision Avoidance 2-6
C. Use of Distractions During Flight Training 2-6
D. Principles of Flight 2-7
E. Elevators, Ailerons, and Rudder 2-7
F. Trim Devices 2-7
G. Wing Flaps 2-8
H. Airplane Weight and Balance 2-8
I. Navigation and Flight Planning 2-8
J. Night Operations 2-9
K. High Altitude Operations 2-10
L. Federal Aviation Regulations 2-10
M. Use of Minimum Equipment List 2-11
N. Publications 2-11
O. National Airspace System 2-12
P. Logbook Entries and Certificate
Endorsements 2-12

CONTENTS (continued)

III. PREFLIGHT PREPARATION

A. Certificates and Documents 2-13
B. Weather Information 2-13
C. Water and Seaplane Characteristics 2-14
D. Seaplane Bases, Rules, and Aids to Marine Navigation 2-14

IV. PREFLIGHT LESSON ON A MANEUVER TO BE PERFORMED IN FLIGHT

Maneuver Lesson 2-15

V. MULTIENGINE OPERATIONS

A. Operation of Systems 2-16
B. Performance and Limitations 2-17
C. Flight Principles - Engine Inoperative 2-18
D. Emergency Procedures 2-19

VI. GROUND AND WATER OPERATIONS

A. Visual Inspection 2-20
B. Cockpit Management 2-21
C. Engine Starting 2-22
D. Taxiing - Landplane 2-23
E. Taxiing - Seaplane 2-24
F. Sailing 2-25
G. Pretakeoff Check 2-26

VII. AIRPORT OPERATIONS

A. Radio Communications and ATC Light Signals 2-27
B. Traffic Patterns 2-28
C. Airport and Runway Markings and Lighting 2-29

VIII. TAKEOFFS AND CLIMBS

A. Normal and Crosswind Takeoff and Climb 2-30
B. Short-Field Takeoff and Climb 2-31
C. Rough-Water Takeoff and Climb 2-33
D. Confined-Area Takeoff and Climb 2-34

CONTENTS (continued)

IX. FUNDAMENTALS OF FLIGHT

A. Straight-and-Level Flight 2-36
B. Level Turns 2-37
C. Straight Climbs and Climbing Turns 2-38
D. Straight Descents and Descending Turns 2-39

X. STALLS AND MANEUVERING DURING SLOW FLIGHT

A. Power-On Stalls 2-40
B. Power-Off Stalls 2-42
C. Maneuvering During Slow Flight 2-43

XI. BASIC INSTRUMENT MANEUVERS

A. Straight-and-Level Flight 2-45
B. Straight, Constant Airspeed Climbs 2-46
C. Straight, Constant Airspeed Descents 2-47
D. Turns to Headings 2-48
E. Recovery from Unusual Flight Attitudes 2-49
F. Radio Aids and Radar Services 2-50

XII. PERFORMANCE MANEUVERS

A. Steep Turns 2-51
B. Emergency Descent 2-52

XIII. GROUND REFERENCE MANEUVERS

A. Rectangular Course 2-53
B. S-Turns Across a Road 2-54
C. Turns Around a Point 2-55

XIV. EMERGENCY OPERATIONS

A. Systems and Equipment Malfunctions 2-56
B. Maneuvering with One Engine Inoperative 2-57
C. Engine Inoperative Loss of Directional Control Demonstration 2-59
D. Demonstrating the Effects of Various Airspeeds and Configurations During Engine Inoperative Performance 2-61

CONTENTS (continued)

E. Engine Failure During Takeoff Before V_{MC} 2-62
F. Engine Failure After Lift-Off 2-63
G. Approach and Landing with an Inoperative Engine 2-64
H. Emergency Equipment and Survival Gear 2-66

XV. APPROACHES AND LANDINGS

A. Normal and Crosswind Approach and Landing 2-67
B. Go-Around 2-68
C. Short-Field Approach and Landing 2-69
D. Glassy-Water Approach and Landing 2-70
E. Rough-Water Approach and Landing 2-72
F. Confined-Area Approach and Landing 2-73

XVI. AFTER-LANDING PROCEDURES

A. Anchoring 2-75
B. Docking and Mooring 2-76
C. Beaching 2-77
D. Ramping 2-78
E. Postflight Procedures 2-78

APPLICANT'S PRACTICAL TEST CHECKLIST

APPOINTMENT WITH INSPECTOR OR EXAMINER:

NAME____________________________________

TIME/DATE_______________________________

ACCEPTABLE AIRCRAFT

- ☐ View-Limiting Device (if applicable)
- ☐ Aircraft Documents:
  - Airworthiness Certificate
  - Registration Certificate
  - Operating Limitations
- ☐ FCC Station License
- ☐ Aircraft Maintenance Records:
  - Airworthiness Inspections
- ☐ Pilot's Operating Handbook, FAA-Approved Airplane Flight Manual

PERSONAL EQUIPMENT

- ☐ Current Aeronautical Charts
- ☐ Computer and Plotter
- ☐ Flight Plan Form
- ☐ Flight Logs
- ☐ Current AIM
- ☐ Current Airport Facility Directory

PERSONAL RECORDS

- ☐ Pilot Certificate
- ☐ Medical Certificate
- ☐ Completed FAA Form 8710-1, Airman Certificate and/or Rating Application
- ☐ AC Form 8080-2, Airman Written Test Report or Computer Test Report
- ☐ Logbook with Instructor's Endorsement
- ☐ Notice of Disapproval (if applicable)
- ☐ Approved School Graduation Certificate (if applicable)
- ☐ Examiner's Fee (if applicable)

EXAMINER'S CHECKLIST
FLIGHT INSTRUCTOR - AIRPLANE
(MULTIENGINE)

APPLICANT'S NAME________________________

EXAMINER'S NAME_________________________

DATE ___________ **TYPE CHECK** ___________

AREA OF OPERATION:

I. FUNDAMENTALS OF INSTRUCTING

- ☐ **A.** The Learning Process
- ☐ **B.** The Teaching Process
- ☐ **C.** Teaching Methods
- ☐ **D.** Evaluation
- ☐ **E.** Flight Instructor Characteristics and Responsibilities
- ☐ **F.** Human Factors
- ☐ **G.** Planning Instructional Activity

II. TECHNICAL SUBJECT AREAS

- ☐ **A.** Aeromedical Factors
- ☐ **B.** Visual Scanning and Collision Avoidance
- ☐ **C.** Use of Distractions During Flight Training
- ☐ **D.** Principles of Flight
- ☐ **E.** Elevators, Ailerons, and Rudder
- ☐ **F.** Trim Devices
- ☐ **G.** Wing Flaps
- ☐ **H.** Airplane Weight and Balance
- ☐ **I.** Navigation and Flight Planning
- ☐ **J.** Night Operations
- ☐ **K.** High Altitude Operations
- ☐ **L.** Federal Aviation Regulations
- ☐ **M.** Use of Minimum Equipment List
- ☐ **N.** Publications
- ☐ **O.** National Airspace System
- ☐ **P.** Logbook Entries and Certificate Endorsements

EXAMINER'S CHECKLIST
FLIGHT INSTRUCTOR-AIRPLANE (MULTIENGINE) (continued)

III. PREFLIGHT PREPARATION

☐ **A.** Certificates and Documents
☐ **B.** Weather Information
☐ **C.** Water and Seaplane Characteristics
☐ **D.** Seaplane Bases, Rules, and Aids to Marine Navigation

IV. PREFLIGHT LESSON ON A MANEUVER TO BE PERFORMED IN FLIGHT

☐ Maneuver Lesson

V. MULTIENGINE OPERATIONS

☐ **A.** Operation of Systems
☐ **B.** Performance and Limitations
☐ **C.** Flight Principles - Engine Inoperative
☐ **D.** Emergency Procedures

VI. GROUND AND WATER OPERATIONS

☐ **A.** Visual Inspection
☐ **B.** Cockpit Management
☐ **C.** Engine Starting
☐ **D.** Taxiing - Landplane
☐ **E.** Taxiing - Seaplane
☐ **F.** Sailing
☐ **G.** Pretakeoff Check

VII. AIRPORT OPERATIONS

☐ **A.** Radio Communications and ATC Light Signals
☐ **B.** Traffic Patterns
☐ **C.** Airport and Runway Markings and Lighting

EXAMINER'S CHECKLIST FLIGHT INSTRUCTOR-AIRPLANE (MULTIENGINE) (continued)

VIII. TAKEOFFS AND CLIMBS

- ☐ **A.** Normal and Crosswind Takeoff and Climb
- ☐ **B.** Short-Field Takeoff and Climb
- ☐ **C.** Rough-Water Takeoff and Climb
- ☐ **D.** Confined-Area Takeoff and Climb

IX. FUNDAMENTALS OF FLIGHT

- ☐ **A.** Straight-and-Level Flight
- ☐ **B.** Level Turns
- ☐ **C.** Straight Climbs and Climbing Turns
- ☐ **D.** Straight Descents and Descending Turns

X. STALLS AND MANEUVERING DURING SLOW FLIGHT

- ☐ **A.** Power-On Stalls
- ☐ **B.** Power-Off Stalls
- ☐ **C.** Maneuvering During Slow Flight

XI. BASIC INSTRUMENT MANEUVERS

- ☐ **A.** Straight-and-Level Flight
- ☐ **B.** Straight, Constant Airspeed Climbs
- ☐ **C.** Straight, Constant Airspeed Descents
- ☐ **D.** Turns to Headings
- ☐ **E.** Recovery from Unusual Flight Attitudes
- ☐ **F.** Radio Aids and Radar Services

XII. PERFORMANCE MANEUVERS

- ☐ **A.** Steep Turns
- ☐ **B.** Emergency Descent

XIII. GROUND REFERENCE MANEUVERS

- ☐ **A.** Rectangular Course
- ☐ **B.** S-Turns Across a Road
- ☐ **C.** Turns Around a Point

EXAMINER'S CHECKLIST FLIGHT INSTRUCTOR-AIRPLANE (MULTIENGINE) (continued)

XIV. EMERGENCY OPERATIONS

- ☐ **A.** Systems and Equipment Malfunctions
- ☐ **B.** Maneuvering with One Engine Inoperative
- ☐ **C.** Engine Inoperative Loss of Directional Control Demonstration
- ☐ **D.** Demonstrating the Effects of Various Airspeeds and Configurations During Engine Inoperative Performance
- ☐ **E.** Engine Failure During Takeoff Before V_{MC}
- ☐ **F.** Engine Failure After Lift-Off
- ☐ **G.** Approach and Landing with an Inoperative Engine
- ☐ **H.** Emergency Equipment and Survival Gear

XV. APPROACHES AND LANDINGS

- ☐ **A.** Normal and Crosswind Approach and Landing
- ☐ **B.** Go-Around
- ☐ **C.** Short-Field Approach and Landing
- ☐ **D.** Glassy-Water Approach and Landing
- ☐ **E.** Rough-Water Approach and Landing
- ☐ **F.** Confined-Area Approach and Landing

XVI. AFTER-LANDING PROCEDURES

- ☐ **A.** Anchoring
- ☐ **B.** Docking and Mooring
- ☐ **C.** Beaching
- ☐ **D.** Ramping
- ☐ **E.** Postflight Procedures

I. AREA OF OPERATION: *FUNDAMENTALS OF INSTRUCTING*

NOTE: The examiner will select at least TASKS E and G.

A. TASK: THE LEARNING PROCESS (AMEL and AMES)

REFERENCE: AC 60-14.

Objective. To determine that the applicant exhibits instructional knowledge of the elements of the learning process by describing:

1. The definition of learning.
2. Characteristics of learning.
3. Practical application of the laws of learning.
4. Factors involved in how people learn.
5. Recognition and proper use of the various levels of learning.
6. Principles that are applied in learning a skill.
7. Factors related to forgetting and retention.
8. How transfer of learning affects the learning process.
9. How the formation of habit patterns affects the learning process.

B. TASK: THE TEACHING PROCESS (AMEL and AMES)

REFERENCE: AC 60-14.

Objective. To determine that the applicant exhibits instructional knowledge of the elements of the teaching process by describing:

1. Preparation for a lesson or an instructional period.
2. Presentation of knowledge and skills, including the methods which are suitable in particular situations.
3. Application, by the student, of the knowledge and skills presented by the instructor.
4. Review of the material presented and the evaluation of student performance and accomplishment.

C. TASK: TEACHING METHODS (AMEL and AMES)

REFERENCE: AC 60-14.

Objective. To determine that the applicant exhibits instructional knowledge of the elements of teaching methods by describing:

1. The organization of a lesson, i.e., introduction, development, and conclusion.
2. The lecture method.
3. The guided discussion method.
4. The demonstration/performance method.
5. Programmed instruction.
6. Audio-visual instruction.

D. TASK: EVALUATION (AMEL and AMES)

REFERENCE: AC 60-14.

Objective. To determine that the applicant exhibits instructional knowledge of the elements of evaluation by describing:

1. The purpose of evaluation.
2. Characteristics of effective oral questions.
3. Types of oral questions to avoid.
4. Responses to student questions.
5. Characteristics and development of effective written tests.
6. Characteristics and uses of performance tests, specifically, the FAA practical test standards.

E. TASK: FLIGHT INSTRUCTOR CHARACTERISTICS AND RESPONSIBILITIES (AMEL and AMES)

REFERENCE: AC 60-14.

Objective. To determine that the applicant exhibits instructional knowledge of the elements of flight instructor characteristics and responsibilities by describing:

1. Major considerations and qualifications which must be included in flight instructor professionalism.
2. Role of the flight instructor in dealing with student stress, anxiety, and psychological abnormalities.
3. Flight instructor's responsibility with regard to student pilot supervision and surveillance.
4. Flight instructor's authority and responsibility for endorsements and recommendations.
5. Flight instructor's responsibility in the conduct of the required FAA flight review.

F. TASK: HUMAN FACTORS (AMEL and AMES)

REFERENCE: AC 60-14.

Objective. To determine that the applicant exhibits instructional knowledge of the elements related to human factors by describing:

1. Control of human behavior.
2. Development of student potential.
3. Relationship of human needs to behavior and learning.
4. Relationship of defense mechanisms to student learning.
5. Relationship of defense mechanisms to pilot decision making.
6. General rules which a flight instructor should follow during student training to ensure good human relations.

G. TASK: PLANNING INSTRUCTIONAL ACTIVITY (AMEL and AMES)

REFERENCE: AC 60-14.

Objective. To determine that the applicant exhibits instructional knowledge of the elements related to the planning of instructional activity by describing:

1. Development of a course of training.
2. Content and use of a training syllabus.
3. Purpose, characteristics, proper use, and items of a lesson plan.
4. Flexibility features of a course of training, syllabus, and lesson plan required to accommodate students with varying backgrounds, levels of experience, and ability.

II. AREA OF OPERATION: *TECHNICAL SUBJECT AREAS*

NOTE: The examiner will select TASK P and at least one other TASK.

A. TASK: AEROMEDICAL FACTORS (AMEL and AMES)

REFERENCES: AC 61-21, AC 67-2; AIM.

Objective. To determine that the applicant exhibits instructional knowledge of the elements related to aeromedical factors by describing:

1. How to obtain an appropriate medical certificate.
2. How to obtain a medical certificate in the event of a possible medical deficiency.[1]
3. Hypoxia, its symptoms, effects, and corrective action.
4. Hyperventilation, its symptoms, effects, and corrective action.
5. Middle ear and sinus problems, their causes, effects, and corrective action.
6. Spatial disorientation, its causes, effects, and corrective action.
7. Motion sickness, its causes, effects, and corrective action.
8. Effects of alcohol and drugs, and their relationship to flight safety.
9. Carbon monoxide poisoning, its symptoms, effects, and corrective action.
10. Effect of nitrogen excesses during scuba dives and how this affects pilots and passengers during flight.
11. Fatigue, its effects and corrective action.

[1]The flight instructor should encourage a person considering flight training to obtain an appropriate medical certificate from an Aviation Medical Examiner before training is started. In the event a person's eligibility to hold a medical certificate is questionable, the flight instructor should be aware that some physical handicaps do not always prohibit activity as pilot of an aircraft. The flight instructor should advise such a person that assistance in obtaining a medical certificate is available through the cooperation of the medical examiner and the local FAA Flight Standards district office. However, this assistance is available only when requested specifically by the person seeking the medical certificate.

B. TASK: VISUAL SCANNING AND COLLISION AVOIDANCE (AMEL and AMES)

REFERENCES: AC 61-21, AC 61-23, AC 67-2, AC 90-48; AIM.

Objective. To determine that the applicant exhibits instructional knowledge of the elements of visual scanning and collision avoidance by describing:

1. Relationship between a pilot's physical or mental condition and vision.
2. Various environmental conditions that degrade vision.
3. Various optical illusions.
4. "See and avoid" concept.
5. Practice of "time sharing" of attention inside and outside the cockpit.
6. Proper visual scanning technique.
7. Relationship between poor visual scanning habits and increased collision risk.
8. Proper clearing procedures.
9. Importance of knowing aircraft blind spots.
10. Relationship between aircraft speed differential and collision risk.
11. Situations which involve the greatest collision risk.

C. TASK: USE OF DISTRACTIONS DURING FLIGHT TRAINING (AMEL and AMES)

REFERENCE: AC 61-92.

Objective. To determine that the applicant exhibits instructional knowledge of the elements of the use of distractions by describing:

1. Flight situations where pilot distraction is a cause factor related to stall/spin accidents.
2. Selection of realistic distractions for specific flight situations.
3. Relationship between division of attention and flight instructor use of distractions.
4. Difference between proper use of distractions and harassment.

D. TASK: PRINCIPLES OF FLIGHT (AMEL and AMES)

REFERENCES: AC 61-21, AC 61-23.

Objective. To determine that the applicant exhibits instructional knowledge of the elements of principles of flight by describing:

1. Airplane and airfoil design characteristics.
2. Forces acting on an airplane in various flight maneuvers.
3. Airplane stability and controllability.
4. Torque effect and correction.
5. Structural integrity and velocity/load factor chart.
6. Wingtip vortices and precautions to be taken.

E. TASK: ELEVATORS, AILERONS, AND RUDDER (AMEL and AMES)

REFERENCES: AC 61-21, AC 61-23.

Objective. To determine that the applicant exhibits instructional knowledge of the elements related to the elevators, ailerons, and rudder by describing:

1. Purpose of each primary control.
2. Location, attachments, and system of control.
3. Direction of movement relative to airflow.
4. Effect on airplane control.
5. Proper technique for use.

F. TASK: TRIM DEVICES (AMEL and AMES)

REFERENCES: AC 61-21, AC 61-23.

Objective. To determine that the applicant exhibits instructional knowledge of the elements related to trim devices by describing:

1. Purpose.
2. Location, attachments, and system of control.
3. Direction of movement relative to airflow and the primary control surface.
4. Effect on airplane control.
5. Proper technique for use.

G. TASK: **WING FLAPS** (AMEL and AMES)

REFERENCES: AC 61-21, AC 61-23.

Objective. To determine that the applicant exhibits instructional knowledge of the elements related to wing flaps by describing:

1. Purpose.
2. Various types.
3. Location, attachments, and system of control.
4. Effect on airplane control.
5. Proper technique for use.

H. TASK: **AIRPLANE WEIGHT AND BALANCE** (AMEL and AMES)

REFERENCES: AC 61-21, AC 61-23, AC 91-23.

Objective. To determine that the applicant exhibits instructional knowledge of the elements of airplane weight and balance by describing:

1. Weight and balance terms.
2. Effect of weight and balance on performance.
3. Methods of weight and balance control.
4. Determination of total weight and center of gravity and the changes that occur when adding, removing, or shifting weight.

I. TASK: **NAVIGATION AND FLIGHT PLANNING** (AMEL and AMES)

REFERENCES: AC 61-21, AC 61-23.

Objective. To determine that the applicant exhibits instructional knowledge of the elements of navigation and flight planning by describing:

1. Terms used in navigation.
2. Features of aeronautical charts.
3. Importance of using the proper and current aeronautical charts.
4. Identification of various types of airspace.

5. Method of plotting a course, selection of fuel stops and alternates, and appropriate actions in the event of unforeseen situations.
6. Fundamentals of pilotage and dead reckoning.
7. Fundamentals of radio navigation.
8. Diversion to an alternate.
9. Lost procedures.
10. Computation of fuel consumption.
11. Importance of preparing and properly using a flight log.
12. Importance of a weather check and the use of good judgment in making a "go/no-go" decision.
13. Purpose of, and procedure used in, filing a flight plan.

J. TASK: NIGHT OPERATIONS (AMEL and AMES)

REFERENCES: AC 61-21, AC 61-23; AIM; FAA-S-8081-1.

Objective. To determine that the applicant exhibits instructional knowledge of the elements of night operations by describing:

1. Factors related to night vision.
2. Disorientation and night optical illusions.
3. Importance of assuring that windshield and windows are clean.
4. Proper adjustment of interior lights.
5. Importance of having a flashlight with a red lens.
6. Night preflight inspection.
7. Engine starting procedures, including use of position and anticollision lights prior to start.
8. Taxiing and orientation on an airport.
9. Takeoff and climb-out.
10. Inflight orientation.
11. Importance of verifying the airplane's attitude by reference to flight instruments.
12. Recovery from critical flight attitudes.
13. Emergencies such as electrical failure, engine malfunction, and emergency landings.
14. Traffic patterns.
15. Approaches and landings with and without landing lights.
16. Go-arounds.

K. TASK: HIGH ALTITUDE OPERATIONS (AMEL and AMES)

REFERENCES: FAR Part 91; AC 61-107, AC 67-2; AIM; Pilot's Operating Handbook, FAA-Approved Airplane Flight Manual.

Objective. To determine that the applicant exhibits instructional knowledge of the elements of high altitude operations by describing:

1. Regulatory requirements for use of oxygen.
2. Physiological hazards associated with high altitude operations.
3. Characteristics of a pressurized airplane and various types of supplemental oxygen systems.
4. Importance of "aviators breathing oxygen."
5. Care and storage of high pressure oxygen bottles.
6. Problems associated with rapid decompression and corresponding solutions.

L. TASK: FEDERAL AVIATION REGULATIONS (AMEL and AMES)

REFERENCES: FAR Parts 61 and 91; NTSB Part 830.

Objective. To determine that the applicant exhibits instructional knowledge of the elements related to Federal Aviation Regulations by describing:

1. Availability and method of revision.
2. FAR Part 61, FAR Part 91, and NTSB Part 830 -

   (a) purpose.
   (b) general content.

M. TASK: USE OF MINIMUM EQUIPMENT LIST (AMEL and AMES)

REFERENCE: FAR Section 91.213.

Objective. To determine that the applicant exhibits instructional knowledge of the elements related to the use of an approved minimum equipment list by describing:

1. Aircraft that require use of a minimum equipment list.
2. Airworthiness limitations imposed on aircraft operations with inoperative instruments or equipment.
3. Requirements for letter of authorization from FAA district office.
4. Supplemental type certificate.
5. Instrument and equipment exemptions.
6. Special flight permit
7. Procedures for deferring maintenance on aircraft without an approved minimum equipment list.

N. TASK: PUBLICATIONS (AMEL and AMES)

REFERENCES: AC 00-2, AC 61-21, AC 61-23; AIM; Pilot's Operating Handbook, FAA-Approved Flight Manual.

Objective. To determine that the applicant exhibits instructional knowledge of the elements related to flight information publications, advisory circulars, practical test standards, pilot operating handbooks, and FAA-approved airplane flight manuals by describing:

1. Availability.
2. Purpose.
3. General content.

O. TASK: NATIONAL AIRSPACE SYSTEM (AMEL and AMES)

REFERENCES: FAR Part 91; AIM.

Objective. To determine that the applicant exhibits instructional knowledge of the elements of the national airspace system by describing:

1. General dimension of airspace segments.
2. Operating limitations associated with uncontrolled, controlled, special use, and other airspace.

P. TASK: LOGBOOK ENTRIES AND CERTIFICATE ENDORSEMENTS (AMEL and AMES)

REFERENCES: FAR Part 61; AC 61-21, AC 61-65.

Objective. To determine that the applicant exhibits instructional knowledge of the elements related to logbook entries and certificate endorsements by describing:

1. Required logbook entries for instruction given.
2. Required student pilot certificate endorsements, including appropriate logbook entries.
3. Preparation of a recommendation for a pilot practical test, including appropriate logbook entry.
4. Required endorsement of a pilot logbook for the satisfactory completion of the required FAA flight review.
5. Required flight instructor records.

III. AREA OF OPERATION: *PREFLIGHT PREPARATION*

NOTE: The examiner will select at least one TASK.

A. TASK: CERTIFICATES AND DOCUMENTS (AMEL and AMES)

REFERENCES: FAR Parts 43, 61, and 91; AC 61-21, AC 61-23; Pilot's Operating Handbook, FAA-Approved Airplane Flight Manual.

Objective. To determine that the applicant exhibits instructional knowledge of the elements related to certificates and documents by describing:

1. Requirements for issuance of pilot and flight instructor certificates and ratings, and the privileges and limitations of those certificates and ratings.
2. Class and duration of medical certificates.
3. Airworthiness and registration certificates.
4. Airplane handbooks and manuals.
5. Airplane maintenance requirements, tests, and records.

B. TASK: WEATHER INFORMATION (AMEL and AMES)

REFERENCES: AC 00-6, AC 00-45, AC 61-21, AC 61-23, AC 61-84.

Objective. To determine that the applicant exhibits instructional knowledge of the elements related to weather information by describing:

1. Importance of a thorough weather check.
2. Various means of obtaining weather information.
3. Use of weather reports, forecasts, and charts.
4. Use of PIREP's, SIGMET's, AIRMET's, and NOTAM's.
5. Recognition of aviation weather hazards to include wind shear.
6. Factors to be considered in making a "go/no-go" decision.

C. TASK: WATER AND SEAPLANE CHARACTERISTICS (AMES)

REFERENCES: AC 61-21; Seaplane Manual.

Objective. To determine that the applicant exhibits instructional knowledge of the elements related to water and seaplane characteristics by describing:

1. Characteristics of water surface as it is affected by such factors as size of the water area, currents, debris, wind, sandbars, islands, or shoals.
2. Seaplane float or hull construction and its relationship to performance.
3. Causes of porpoising and skipping and pilot action necessary to prevent or to correct those occurrences.

D. TASK: SEAPLANE BASES, RULES, AND AIDS TO MARINE NAVIGATION (AMES)

REFERENCES: FAR Part 91; AC 61-21; Rules of the Road; Aids to Marine Navigation.

Objective. To determine that the applicant exhibits instructional knowledge of the elements related to seaplane bases, rules, and aids to marine navigation by describing:

1. How to locate and identify seaplane bases on charts or in directories.
2. Operating restrictions at various seaplane bases.
3. Right-of-way, steering, and sailing rules pertinent to seaplane operation.
4. Purpose and identification of marine navigation aids such as buoys, beacons, lights, and sound signals.

IV. AREA OF OPERATION: *PREFLIGHT LESSON ON A MANEUVER TO BE PERFORMED IN FLIGHT*

NOTE: Examiner will select at least one maneuver from AREAS OF OPERATION VII through XV, and ask the applicant to present a preflight lesson on the selected maneuver as the lesson would be taught to a student.

TASK: MANEUVER LESSON (AMEL and AMES)

REFERENCES: AC 60-14, AC 61-21, AC 61-23; FAA-S-8081-1, FAA-S-8081-2; Pilot's Operating Handbook, FAA-Approved Airplane Flight Manual; Seaplane Manual.

Objective. To determine that the applicant exhibits instructional knowledge of the selected maneuver by:

1. Stating the purpose.
2. Giving an accurate, comprehensive oral description, including the elements and common errors.
3. Using instructional aids, as appropriate.
4. Describing the recognition, analysis, and correction of common errors.

V. AREA OF OPERATION: *MULTIENGINE OPERATIONS*

NOTE: The examiner will select TASKS A, B, C, and D.

A. TASK: OPERATION OF SYSTEMS (AMEL and AMES)

REFERENCES: AC 61-21; FAA-S-8081-1, FAA-S-8081-2; Pilot's Operating Handbook, FAA-Approved Airplane Flight Manual.

Objective. To determine that the applicant exhibits instructional knowledge of the elements related to the operation of systems, as applicable to the airplane used for the practical test, by describing:

1. Primary flight controls and trim.
2. Pitot static/vacuum system and associated instruments.
3. Landing gear.
4. Wing flaps, leading edge devices, and spoilers.
5. Powerplant, including controls, indicators, cooling, and fire detection.
6. Propellers, including controls and indicators.
7. Fuel, oil, and hydraulic systems.
8. Electrical system.
9. Environmental system.
10. Deicing and anti-icing systems.
11. Avionics system.
12. Any system unique to the airplane flown.

B. TASK: PERFORMANCE AND LIMITATIONS (AMEL and AMES)

REFERENCES: AC 61-21, AC 61-23, AC 61-84, AC 91-23; FAA-S-8081-1, FAA-S-8081-2; Pilot's Operating Handbook, FAA-Approved Airplane Flight Manual.

Objective. To determine that the applicant exhibits instructional knowledge of the elements related to performance and limitations by describing:

1. Determination of weight and balance condition.
2. Use of performance charts, tables, and other data in determining performance in various phases of flight.
3. Effects of exceeding limitations.
4. Effects of atmospheric conditions on performance.
5. Factors to be considered in determining that the required performance is within the airplane's capabilities.

C. TASK: FLIGHT PRINCIPLES - ENGINE INOPERATIVE (AMEL and AMES)

REFERENCES: AC 61-21; FAA-S-8081-1, FAA-S-8081-2; Pilot's Operating Handbook, FAA-Approved Airplane Flight Manual.

Objective. To determine that the applicant exhibits instructional knowledge of the elements related to flight principles - engine inoperative by describing:

1. Effects of density altitude.
2. Importance of reducing drag.
3. Importance of establishing and maintaining proper airspeed.
4. Importance of maintaining proper pitch and bank attitudes, and proper coordination of controls.
5. Effects of weight and center of gravity.
6. Critical engine.
7. Reasons for loss of directional control.
8. Indications of approaching loss of directional control.
9. Reasons for variations in V_{MC}.
10. Relationship of V_{MC} to stall speed, including determination of whether a loss of directional control demonstration can be safely accomplished.
11. Takeoff emergencies, including planning, decisions, and single-engine operations.

D. TASK: EMERGENCY PROCEDURES (AMEL and AMES)

REFERENCES: AC 61-21; FAA-S-8081-1, FAA-S-8081-2; Pilot's Operating Handbook, FAA-Approved Airplane Flight Manual.

Objective. To determine that the applicant exhibits instructional knowledge of the elements of emergency procedures appropriate to the airplane used for the practical test by describing:

1. Importance of availability and use of an emergency checklist.
2. Possible causes of partial or complete power loss in various flight situations.
3. Procedures to be followed if partial or complete power loss occurs during any phase of flight.
4. Procedures to be followed in icing conditions.
5. Procedures to be followed in the event of instrument malfunction.
6. Recommended recovery procedure from an unintentional spin.
7. Any other emergency appropriate to the airplane flown.

VI. AREA OF OPERATION: *GROUND AND WATER OPERATIONS*

NOTE: The examiner will select at least one TASK.

A. TASK: VISUAL INSPECTION (AMEL and AMES)

REFERENCES: AC 60-14, AC 61-21, FAA-S-8081-1, FAA-S-8081-2; Pilot's Operating Handbook, FAA-Approved Airplane Flight Manual.

Objective. To determine that the applicant:

1. Exhibits instructional knowledge of the elements of a visual inspection, as applicable to the airplane used for the practical test, by describing -

   (a) reasons for the visual inspection, items that should be inspected, and how defects are detected.
   (b) importance of using the appropriate checklist.
   (c) how to determine fuel and oil quantity.
   (d) methods used to determine fuel and oil contamination.
   (e) detection of fuel, oil, and hydraulic leaks.
   (f) inspection of the oxygen system, including supply and proper operation.
   (g) inspection of the flight controls and water rudder.
   (h) detection of visible structural damage.
   (i) removal of tie-downs, control locks, and wheel chocks.
   (j) removal of ice and frost.
   (k) importance of proper loading and securing of baggage, cargo, and equipment.
   (l) use of sound judgment in determining whether the airplane is in condition for safe flight.

2. Exhibits instructional knowledge of common errors related to a visual inspection by describing -

   (a) failure to use or improper use of checklist.
   (b) hazards which may result from allowing distractions to interrupt a visual inspection.
   (c) inability to recognize discrepancies.
   (d) failure to assure servicing with the proper fuel and oil.

3. Demonstrates and simultaneously explains a visual inspection from an instructional standpoint.

B. TASK: COCKPIT MANAGEMENT (AMEL and AMES)

REFERENCES: AC 60-14, AC 61-21; FAA-S-8081-1, FAA-S-8081-2; Pilot's Operating Handbook, FAA-Approved Airplane Flight Manual.

Objective. To determine that the applicant:

1. Exhibits instructional knowledge of the elements of cockpit management by describing -

   (a) proper arranging and securing of essential materials and equipment in the cockpit.
   (b) proper and orderly maintenance of records that reflect the progress of the flight.
   (c) proper use and/or adjustment of such cockpit items as safety belts, shoulder harnesses, rudder pedals, and seats.
   (d) occupant briefing on emergency procedures and use of safety belts.

2. Exhibits instructional knowledge of common errors related to cockpit management by describing -

   (a) failure to place and secure essential materials and equipment for easy access during flight.
   (b) failure to maintain accurate records essential to the progress of the flight.
   (c) improper adjustment of equipment and controls.

3. Demonstrates and simultaneously explains cockpit management from an instructional standpoint.

C. TASK: **ENGINE STARTING** (AMEL and AMES)

REFERENCES: AC 60-14, AC 61-21, AC 61-23, AC 91-13, AC 91-55; FAA-S-8081-1, FAA-S-8081-2; Pilot's Operating Handbook, FAA-Approved Airplane Flight Manual.

Objective. To determine that the applicant:

1. Exhibits instructional knowledge of the elements of engine starting, as applicable to the airplane used for the practical test by describing -

   (a) safety precautions related to starting.
   (b) use of external power.
   (c) effect of atmospheric conditions on starting.
   (d) importance of following the appropriate checklist.
   (e) adjustment of engine controls during start.
   (f) prevention of airplane movement during and after start.

2. Exhibits instructional knowledge of common errors related to engine starting by describing -

   (a) failure to use or the improper use of the checklist.
   (b) excessively high RPM after starting.
   (c) improper preheat of the engine during severe cold weather conditions.
   (d) failure to assure proper clearance of the propellers.

3. Demonstrates and simultaneously explains engine starting from an instructional standpoint.

D. TASK: TAXIING - LANDPLANE (AMEL)

REFERENCES: AC 60-14, AC 61-21; FAA-S-8081-1, FAA-S-8081-2; Pilot's Operating Handbook, FAA-Approved Airplane Flight Manual.

Objective. To determine that the applicant:

1. Exhibits instructional knowledge of the elements of landplane taxiing by describing -

   (a) proper brake check and correct use of brakes.
   (b) compliance with airport surface marking, signals, and clearances.
   (c) how to control direction and speed.
   (d) control positioning for various wind conditions.
   (e) techniques used to avoid other aircraft and hazards.

2. Exhibits instructional knowledge of common errors related to landplane taxiing by describing -

   (a) improper use of brakes.
   (b) improper positioning of the flight controls for various wind conditions.
   (c) hazards of taxiing too fast.
   (d) failure to comply with markings, signals, or clearances.

3. Demonstrates and simultaneously explains landplane taxiing from an instructional standpoint.
4. Analyzes and corrects simulated common errors related to landplane taxiing.

E. TASK: **TAXIING - SEAPLANE** (AMES)

REFERENCES: AC 60-14, AC 61-21; FAA-S-8081-1, FAA-S-8081-2; Seaplane Manual.

Objective. To determine that the applicant:

1. Exhibits instructional knowledge of the elements of seaplane taxiing by describing -

   (a) wind effect.
   (b) prevention of porpoising and skipping.
   (c) selection of the most suitable course for taxiing, following available marking aids.
   (d) conditions where idle, plowing, and step taxiing are used.
   (e) techniques for idle, plowing, and step taxiing.
   (f) control positioning for various wind conditions.
   (g) use of water rudders.
   (h) techniques used to avoid other aircraft and hazards.
   (i) techniques used to avoid excessive water spray into the propeller.
   (j) 180° and 360° turns in idle, plowing, and step positions.
   (k) application of right-of-way rules.

2. Exhibits instructional knowledge of common errors related to seaplane taxiing by describing -

   (a) improper positioning of flight controls for various wind conditions.
   (b) improper control of speed and direction.
   (c) failure to prevent porpoising and skipping.
   (d) failure to use the most suitable course and available marking aids.
   (e) failure to use proper clearing procedures to avoid hazards.
   (f) failure to apply right-of-way rules.

3. Demonstrates and simultaneously explains seaplane taxiing from an instructional standpoint.
4. Analyzes and corrects simulated common errors related to seaplane taxiing.

F. TASK: SAILING (AMES)

REFERENCES: AC 60-14, AC 61-21; FAA-S-8081-1, FAA-S-8081-2; Pilot's Operating Handbook, FAA-Approved Airplane Flight Manual.

Objective. To determine that the applicant:

1. Exhibits instructional knowledge of the elements of sailing by describing -

   (a) techniques used in sailing (engines idling or shut down, as appropriate).
   (b) conditions and situations where sailing would be used.
   (c) selection of the most favorable course to follow.
   (d) use of flight controls, flaps, doors, and water rudders to follow the selected course.
   (e) techniques used to change direction from downwind to crosswind.
   (f) control of speed.

2. Exhibits instructional knowledge of common errors related to sailing by describing -

   (a) failure to select the most favorable course to destination.
   (b) improper use of controls, flaps, and water rudders.
   (c) improper technique when changing direction.
   (d) improper technique for speed control.

3. Demonstrates and simultaneously explains sailing from an instructional standpoint.
4. Analyzes and corrects simulated common errors related to sailing.

G. TASK: **PRETAKEOFF CHECK** (AMEL and AMES)

REFERENCES: AC 60-14, AC 61-21; FAA-S-8081-1, FAA-S-8081-2; Pilot's Operating Handbook, FAA-Approved Airplane Flight Manual.

Objective. To determine that the applicant:

1. Exhibits instructional knowledge of the elements of a pretakeoff check by describing -

   (a) positioning the airplane with consideration for other aircraft, surface conditions, and wind.
   (b) division of attention inside and outside the cockpit.
   (c) importance of following the checklist and responding to each checklist item.
   (d) reasons for ensuring suitable engine temperatures and pressures for run-up and takeoff.
   (e) method used to determine that airplane is in a safe operating condition.
   (f) importance of reviewing takeoff performance airspeeds, expected takeoff distances, and emergency procedures.
   (g) method used for assuring that the takeoff area or path is free of hazards.
   (h) method used for assuring adequate clearance from other traffic.

2. Exhibits instructional knowledge of common errors related to a pretakeoff check by describing -

   (a) failure to use or the improper use of the checklist.
   (b) improper positioning of the airplane.
   (c) acceptance of marginal engine performance.
   (d) an improper check of flight controls.
   (e) hazards of failure to review takeoff and emergency procedures.
   (f) failure to check for hazards and other traffic.

3. Demonstrates and simultaneously explains a pretakeoff check from an instructional standpoint.
4. Analyzes and corrects simulated common errors related to a pretakeoff check.

VII. AREA OF OPERATION: *AIRPORT OPERATIONS*

NOTE: The examiner will select at least one TASK.

A. TASK: RADIO COMMUNICATIONS AND ATC LIGHT SIGNALS (AMEL and AMES)

REFERENCES: AC 61-21, AC 61-23; FAA-S-8081-1, FAA-S-8081-2.

Objective. To determine that the applicant:

1. Exhibits instructional knowledge of the elements of radio communications and ATC light signals by describing -

   (a) selection and use of appropriate radio frequencies.
   (b) recommended procedure and phraseology for radio voice communications.
   (c) receipt, acknowledgement of, and compliance with, ATC clearances and other instructions.
   (d) interpretation of, and compliance with, ATC light signals.

2. Exhibits instructional knowledge of common errors related to radio communications and ATC light signals by describing -

   (a) use of improper frequencies.
   (b) improper techniques and phraseologies when transmitting radio voice communications.
   (c) failure to acknowledge, or properly comply with, ATC clearances and other instructions.
   (d) failure to understand, or to properly comply with, ATC light signals.

B. TASK: TRAFFIC PATTERNS (AMEL and AMES)

REFERENCES: AC 60-14, AC 61-21, AC 61-23; AIM; FAA-S-8081-1, FAA-S-8081-2.

Objective. To determine that the applicant:

1. Exhibits instructional knowledge of the elements of traffic patterns by describing -

   (a) operations at controlled and uncontrolled airports and seaplane bases.
   (b) adherence to traffic pattern procedures, instructions, and rules.
   (c) how to maintain proper spacing from other traffic.
   (d) how to maintain the proper ground track.
   (e) wind shear and wake turbulence.
   (f) orientation with the runway or landing area in use.
   (g) how to establish a final approach at an appropriate distance from approach end of the runway or landing area.
   (h) use of checklist.

2. Exhibits instructional knowledge of common errors related to traffic patterns by describing -

   (a) failure to comply with traffic pattern instructions, procedures, and rules.
   (b) improper correction for wind drift.
   (c) improper spacing from other traffic.
   (d) poor altitude or airspeed control.

3. Demonstrates and simultaneously explains traffic patterns from an instructional standpoint.
4. Analyzes and corrects simulated common errors related to traffic patterns.

C. TASK: AIRPORT AND RUNWAY MARKINGS AND LIGHTING (AMEL and AMES)

REFERENCES: AC 61-21, AC 61-23; AIM; FAA-S-8081-1, FAA-S-8081-2.

Objective. To determine that the applicant exhibits instructional knowledge of the elements of airport and runway markings and lighting by describing:

1. Identification and proper interpretation of airport, runway, and taxiway markings.
2. Identification and proper interpretation of airport, runway, and taxiway lighting.

VIII. AREA OF OPERATION: *TAKEOFFS AND CLIMBS*

NOTE: The examiner will select at least one TASK.

A. TASK: NORMAL AND CROSSWIND TAKEOFF AND CLIMB (AMEL and AMES)

REFERENCES: AC 60-14, AC 61-21; FAA-S-8081-1, FAA-S-8081-2; Pilot's Operating Handbook, FAA-Approved Airplane Flight Manual, Seaplane Manual.

Objective. To determine that the applicant:

1. Exhibits instructional knowledge of the elements of a normal and crosswind takeoff and climb by describing -

   (a) review of wind conditions.
   (b) takeoff hazards.
   (c) use of wing flaps.
   (d) alignment with takeoff path.
   (e) initial positioning of flight controls.
   (f) power application.
   (g) directional control during acceleration on the surface.
   (h) crosswind control technique during acceleration on the surface.
   (i) rotation at airspeed appropriate for the airplane flown.
   (j) how to establish the single-engine best rate-of-climb pitch attitude.
   (k) how to establish and maintain V_Y.
   (l) how to establish and maintain cruise climb airspeed and the appropriate power setting.
   (m) crosswind correction and track during climb.
   (n) use of checklist.
   (o) difference between a normal and a glassy-water takeoff (seaplane).

2. Exhibits instructional knowledge of common errors related to a normal and crosswind takeoff and climb by describing -

   (a) improper initial positioning of flight controls or wing flaps.
   (b) improper power application.
   (c) inappropriate removal of hand from throttles.
   (d) poor directional control.
   (e) improper use of ailerons.
   (f) rotation at improper airspeed.
   (g) failure to establish and maintain proper climb configuration and airspeeds.
   (h) drift during climb.

3. Demonstrates and simultaneously explains a normal or a crosswind takeoff and climb from an instructional standpoint.
4. Analyzes and corrects simulated common errors related to a normal or a crosswind takeoff and climb.

B. TASK: SHORT-FIELD TAKEOFF AND CLIMB (AMEL)

REFERENCES: AC 60-14, AC 61-21; FAA-S-8081-1, FAA-S-8081-2; Pilot's Operating Handbook, FAA-Approved Airplane Flight Manual.

Objective. To determine that the applicant:

1. Exhibits instructional knowledge of the elements of a short-field takeoff and climb by describing -

   (a) review of wind conditions.
   (b) takeoff and climb hazards, particularly those related to obstacles.
   (c) use of wing flaps.
   (d) how to position and align the airplane for maximum utilization of available takeoff area.
   (e) initial positioning of flight controls.
   (f) power application.
   (g) directional control during acceleration on the surface.
   (h) rotation at the airspeed appropriate for airplane used in the practical test.
   (i) initial climb attitude and airspeed (V_X) until obstacle is cleared (50 feet/16 meters AGL).

(j) acceleration to V_Y and establishment and maintenance of V_Y.
(k) how to establish and maintain cruise climb airspeed and the appropriate power setting.
(l) track during climb.
(m) use of checklist.

2. Exhibits instructional knowledge of common errors related to a short-field takeoff and climb by describing -

(a) failure to position the airplane for maximum utilization of available takeoff area.
(b) improper initial positioning of flight controls and wing flaps.
(c) improper power application.
(d) inappropriate removal of hand from throttles.
(e) poor directional control.
(f) improper use of ailerons.
(g) improper use of brakes.
(h) rotation at improper airspeed.
(i) failure to establish and maintain proper climb configuration and airspeeds.
(j) drift during climb.

3. Demonstrates and simultaneously explains a short-field takeoff and climb from an instructional standpoint.
4. Analyzes and corrects simulated common errors related to a short-field takeoff and climb.

C. TASK: ROUGH-WATER TAKEOFF AND CLIMB (AMES)

REFERENCES: AC 60-14, AC 61-21; FAA-S-8081-1, FAA-S-8081-2; Pilot's Operating Handbook, FAA-Approved Airplane Flight Manual, Seaplane Manual.

Objective. To determine that the applicant:

1. Exhibits instructional knowledge of the elements of a rough-water takeoff and climb by describing -

    (a) review of wind conditions.
    (b) takeoff hazards.
    (c) factors which should be considered in the selection of the most suitable takeoff area.
    (d) use of wing flaps.
    (e) alignment with proposed takeoff path.
    (f) initial positioning of flight controls.
    (g) power application.
    (h) directional control.
    (i) most efficient planing angle.
    (j) rotation at the airspeed appropriate to the airplane used in the practical test.
    (k) how to establish the single-engine best rate-of-climb pitch attitude and acceleration to V_Y.
    (l) how to establish and maintain V_Y.
    (m) how to establish and maintain cruise climb airspeed and the appropriate power setting.
    (n) track during climb.
    (o) use of checklist.

2. Exhibits instructional knowledge of common errors related to a rough-water takeoff and climb by describing -

    (a) poor judgment in the selection of a suitable takeoff area.
    (b) improper initial positioning of flight controls and wing flaps.
    (c) improper power application.
    (d) inappropriate removal of hand from throttles.
    (e) poor directional control.
    (f) improper correction for porpoising, skipping, or an unusual increase in water drag.
    (g) improper pitch attitude during takeoff run and lift-off.

(h) hazards of inadvertent contact with the water after becoming airborne.
(i) failure to establish and maintain proper climb configuration and airspeeds.
(j) drift during climb.

3. Demonstrates and simultaneously explains a rough water takeoff and climb from an instructional standpoint.
4. Analyzes and corrects simulated common errors related to a rough-water takeoff and climb.

D. TASK: CONFINED-AREA TAKEOFF AND CLIMB (AMES)

REFERENCES: AC 60-14, AC 61-21; FAA-S-8081-1, FAA-S-8081-2; Pilot's Operating Handbook, FAA-Approved Airplane Flight Manual, Seaplane Manual.

Objective. To determine that the applicant:

1. Exhibits instructional knowledge of the elements of a confined-area takeoff and climb by describing -

(a) review of wind conditions.
(b) takeoff hazards, particularly those related to obstacles.
(c) use of wing flaps.
(d) factors related to minimum takeoff run and maximum climb performance.
(e) factors that should be considered in the selection of the most favorable takeoff area.
(f) alignment with the proposed takeoff path.
(g) initial positioning of flight controls.
(h) power application.
(i) directional control during acceleration on the surface.
(j) the most efficient planing angle.
(k) rotation at the airspeed appropriate for airplane used for the practical test.
(l) initial climb attitude and airspeed (V_X) until obstacle is cleared (50 feet/16 meters AGL).
(m) acceleration to V_Y and the establishment and maintenance of V_Y.

(n) how to establish and maintain cruise climb airspeed and the appropriate power setting.
(o) track during climb.
(p) use of checklist.

2. Exhibits instructional knowledge of common errors related to a confined-area takeoff and climb by describing -

(a) failure to position the airplane to take advantage of the available takeoff area.
(b) improper initial positioning of flight controls or wing flaps.
(c) improper power application.
(d) inappropriate removal of hand from throttles.
(e) poor directional control.
(f) rotation of improper airspeed.
(g) improper correction for porpoising, skipping, or an unusual increase in water drag.
(h) hazards of inadvertent contact with the water after becoming airborne.
(i) failure to establish and maintain proper climb configuration and airspeeds.
(j) drift during climb.

3. Demonstrates and simultaneously explains a confined-area takeoff and climb from an instructional standpoint.
4. Analyzes and corrects simulated common errors related to a confined-area takeoff and climb.

IX. AREA OF OPERATION: *FUNDAMENTALS OF FLIGHT*

NOTE: The examiner will select at least one TASK.

A. TASK: STRAIGHT-AND-LEVEL FLIGHT (AMEL and AMES)

REFERENCES: AC 60-14, AC 61-21.

Objective. To determine that the applicant:

1. Exhibits instructional knowledge of the elements of straight-and-level flight by describing -

   (a) effect and use of flight controls.
   (b) the Integrated Flight Instruction method.
   (c) outside and instrument references used for pitch, bank, and power control; the cross-check and interpretation of those references; and the control technique used.
   (d) trim technique.
   (e) methods that can be used to overcome tenseness and overcontrolling.

2. Exhibits instructional knowledge of common errors related to straight-and-level flight by describing -

   (a) failure to cross-check and correctly interpret outside and instrument references.
   (b) application of control movements rather than pressures.
   (c) uncoordinated use of flight controls.
   (d) faulty trim technique.

3. Demonstrates and simultaneously explains straight-and-level flight from an instructional standpoint.
4. Analyzes and corrects simulated common errors related to straight-and-level flight.

B. TASK: LEVEL TURNS (AMEL and AMES)

REFERENCES: AC 60-14, AC 61-21.

Objective. To determine that the applicant:

1. Exhibits instructional knowledge of the elements of level turns by describing -

    (a) effect and use of flight controls.
    (b) the Integrated Flight Instruction method.
    (c) outside and instrument references used for pitch, bank, and power control; the cross-check and interpretation of those references; and the control technique used.
    (d) trim technique.
    (e) methods that can be used to overcome tenseness and overcontrolling.

2. Exhibits instructional knowledge of common errors related to level turns by describing -

    (a) failure to cross-check and correctly interpret outside and instrument references.
    (b) application of control movements rather than pressures.
    (c) uncoordinated use of flight controls.
    (d) faulty altitude and bank control.

3. Demonstrates and simultaneously explains a level turn from an instructional standpoint.
4. Analyzes and corrects simulated common errors related to level turns.

C. TASK: **STRAIGHT CLIMBS AND CLIMBING TURNS** (AMEL and AMES)

REFERENCES: AC 60-14, AC 61-21.

Objective. To determine that the applicant:

1. Exhibits instructional knowledge of the elements of straight climbs and climbing turns by describing -

   (a) effect and use of flight controls.
   (b) the Integrated Flight Instruction method.
   (c) outside and instrument references used for pitch, bank, and power control; the cross-check and interpretation of those references; and the control technique used.
   (d) trim technique.
   (e) methods that can be used to overcome tenseness and overcontrolling.

2. Exhibits instructional knowledge of common errors related to straight climbs and climbing turns by describing -

   (a) failure to cross-check and correctly interpret outside and instrument references.
   (b) application of control movements rather than pressures.
   (c) improper correction for torque effect.
   (d) faulty trim technique.

3. Demonstrates and simultaneously explains a straight climb and a climbing turn from an instructional standpoint.
4. Analyzes and corrects simulated common errors related to straight climbs and climbing turns.

D. TASK: STRAIGHT DESCENTS AND DESCENDING TURNS (AMEL and AMES)

REFERENCES: AC 60-14, AC 61-21.

Objective. To determine that the applicant:

1. Exhibits instructional knowledge of the elements of straight descents and descending turns by describing -

   (a) effect and use of flight controls.
   (b) the Integrated Flight Instruction method.
   (c) outside and instrument references used for pitch, bank, and power control; the cross-check and interpretation of those references; and the control technique used.
   (d) trim technique.
   (e) methods that can be used to overcome tenseness and overcontrolling.

2. Exhibits instructional knowledge of common errors related to straight descents and descending turns by describing -

   (a) failure to cross-check and correctly interpret outside and instrument references.
   (b) application of control movements rather than pressures.
   (c) uncoordinated use of flight controls.
   (d) faulty trim technique.
   (e) failure to clear engine and use carburetor heat, as appropriate.

3. Demonstrates and simultaneously explains a straight descent and a descending turn from an instructional standpoint.
4. Analyzes and corrects simulated common errors related to straight descents and descending turns.

AREA OF OPERATION: *STALLS AND MANEUVERING DURING SLOW FLIGHT*

NOTE: The examiner will select at least one TASK. Stalls will not be performed with one engine at reduced power or inoperative and the other engine(s) developing effective power.

Stalls using high power settings should not be performed. The high pitch angles necessary to induce these stalls could possibly result in uncontrollable flight.

Examiners and instructors should be alert to the possible development of high sink rates when performing stalls in multiengine airplanes with high wing loadings; therefore, the altitude loss during stall entries should be no more than 50 feet.

A. TASK: POWER-ON STALLS (AMEL and AMES)

REFERENCES: AC 60-14, AC 61-21; FAA-S-8081-1, FAA-S-8081-2; Pilot's Operating Handbook, FAA-Approved Airplane Flight Manual.

Objective. To determine that the applicant:

1. Exhibits instructional knowledge of the elements of power-on stalls, in climbing flight (straight or turning), with selected landing gear and flap configurations, by describing -

   (a) aerodynamics of power-on stalls.
   (b) relationship of various factors such as landing gear and flap configuration, weight, center of gravity, load factor, and bank angle to stall speed.
   (c) flight situations where unintentional power-on stalls may occur.
   (d) recognition of the first indications of power-on stalls.
   (e) performance of power-on stalls in climbing flight (straight or turning).
   (f) entry technique and minimum entry altitude.
   (g) coordination of flight controls.
   (h) recovery technique and minimum recovery altitude.

2. Exhibits instructional knowledge of common errors related to power-on stalls, in climbing flight (straight or turning), with selected landing gear and flap configurations, by describing -

    (a) failure to establish the specified landing gear and flap configuration prior to entry.
    (b) improper pitch, heading, and bank control during straight ahead stalls.
    (c) improper pitch and bank control during turning stalls.
    (d) rough or uncoordinated control technique.
    (e) failure to recognize the first indications of a stall.
    (f) failure to achieve a stall.
    (g) improper torque correction.
    (h) poor stall recognition and delayed recovery.
    (i) excessive altitude loss or excessive airspeed during recovery.
    (j) secondary stall during recovery.

3. Demonstrates and simultaneously explains power-on stalls, in climbing flight (straight or turning), with selected landing gear and flap configurations, from an instructional standpoint.
4. Analyzes and corrects simulated common errors related to power-on stalls, in climbing flight (straight and turning), with selected landing gear and flap configurations.

B. TASK: POWER-OFF STALLS (AMEL and AMES)

REFERENCES: AC 60-14, AC 61-21; FAA-S-8081-1, FAA-S-8081-2; Pilot's Operating Handbook, FAA-Approved Airplane Flight Manual.

Objective. To determine that the applicant:

1. Exhibits instructional knowledge of the elements of power-off stalls, in descending flight (straight or turning), with selected landing gear and flap configurations, by describing -

   (a) aerodynamics of power-off stalls.
   (b) relationship of various factors such as landing gear and flap configuration, weight, center of gravity, load factor, and bank angle to stall speed.
   (c) flight situations where unintentional power-off stalls may occur.
   (d) recognition of the first indications of power-off stalls.
   (e) performance of power-off stalls in descending flight (straight or turning).
   (f) entry technique and minimum entry altitude.
   (g) coordination of flight controls.
   (h) recovery technique and minimum recovery altitude.

2. Exhibits instructional knowledge of common errors related to power-off stalls, in descending flight (straight or turning), with selected landing gear and flap configurations, by describing -

   (a) failure to establish the specified landing gear and flap configuration prior to entry.
   (b) improper pitch, heading, and bank control during straight ahead stalls.
   (c) improper pitch and bank control during turning stalls.
   (d) rough or uncoordinated control technique.
   (e) failure to recognize the first indications of a stall.
   (f) failure to achieve a stall.
   (g) improper torque correction.

(h) poor stall recognition and delayed recovery.
(i) excessive altitude loss or excessive airspeed during recovery.
(j) secondary stall during recovery.

3. Demonstrates and simultaneously explains power-off stalls, in descending flight (straight or turning), with selected landing gear and flap configurations, from an instructional standpoint.
4. Analyzes and corrects simulated common errors related to power-off stalls, in descending flight (straight or turning), with selected landing gear and flap configurations.

C. TASK: MANEUVERING DURING SLOW FLIGHT (AMEL and AMES)

REFERENCES: AC 60-14, AC 61-21; FAA-S-8081-1, FAA-S-8081-2; Pilot's Operating Handbook, FAA-Approved Airplane Flight Manual.

Objective. To determine that the applicant:

1. Exhibits instructional knowledge of the elements of maneuvering during slow flight by describing -

(a) relationship of configuration, weight, center of gravity, maneuvering loads, angle of bank, and power to flight characteristics and controllability.
(b) relationship of the maneuver to critical flight situations, such as go-arounds.
(c) performance of the maneuver with selected landing gear and flap configurations in straight-and-level flight and level turns.
(d) specified airspeed for the maneuver.
(e) coordination of flight controls.
(f) trim technique.
(g) re-establishment of cruise flight.

2. Exhibits instructional knowledge of common errors related to maneuvering during slow flight by describing -

   (a) failure to establish specified gear and flap configuration.
   (b) improper entry technique.
   (c) failure to establish and maintain the specified airspeed.
   (d) excessive variations of altitude and heading when a constant altitude and heading are specified.
   (e) rough or uncoordinated control technique.
   (f) improper correction for torque effect.
   (g) improper trim technique.
   (h) unintentional stalls.
   (i) inappropriate removal of hand from throttles.

3. Demonstrates and simultaneously explains maneuvering during slow flight from an instructional standpoint.
4. Analyzes and corrects simulated common errors related to maneuvering during slow flight.

XI. AREA OF OPERATION: *BASIC INSTRUMENT MANEUVERS*

NOTE: The examiner will select at least one TASK.

A. TASK: STRAIGHT-AND-LEVEL FLIGHT (AMEL and AMES)

REFERENCES: AC 60-14, AC 61-21, AC 61-27; FAA-S-8081-1.

Objective. To determine that the applicant:

1. Exhibits instructional knowledge of the elements of straight-and-level flight, solely by reference to instruments, by describing -

   (a) instrument cross-check, instrument interpretation, and aircraft control.
   (b) instruments used for pitch, bank, and power control, and how those instruments are used to maintain altitude, heading, and airspeed.
   (c) trim technique.

2. Exhibits instructional knowledge of common errors related to straight-and-level flight solely by reference to instruments by describing -

   (a) "fixation," "omission," and "emphasis" errors during instrument cross-check.
   (b) improper instrument interpretation.
   (c) improper control applications.
   (d) failure to establish proper pitch, bank, or power adjustments during altitude, heading, or airspeed corrections.
   (e) faulty trim technique.

3. Demonstrates and simultaneously explains straight-and-level flight, solely by reference to instruments, from an instructional standpoint.
4. Analyzes and corrects simulated common errors related to straight-and-level flight, solely by reference to instruments.

B. TASK: STRAIGHT, CONSTANT AIRSPEED CLIMBS (AMEL and AMES)

REFERENCES: AC 60-14, AC 61-21, AC 61-27; FAA-S-8081-1.

Objective. To determine that the applicant:

1. Exhibits instructional knowledge of the elements of straight, constant airspeed climbs, solely by reference to instruments, by describing -

   (a) instrument cross-check, instrument interpretation, and aircraft control.
   (b) instruments used for pitch, bank, and power control during entry, during the climb, and during level-off, and how those instruments are used to maintain climb heading and airspeed.
   (c) trim technique.

2. Exhibits instructional knowledge of common errors related to straight, constant airspeed climbs, solely by reference to instruments, by describing -

   (a) "fixation," "omission," and "emphasis" errors during instrument cross-check.
   (b) improper instrument interpretation.
   (c) improper control applications.
   (d) failure to establish proper pitch, bank, or power adjustments during heading and airspeed corrections.
   (e) improper entry or level-off technique.
   (f) faulty trim technique.

3. Demonstrates and simultaneously explains a straight, constant airspeed climb, solely by reference to instruments, from an instructional standpoint.
4. Analyzes and corrects simulated common errors related to straight, constant airspeed climbs, solely by reference to instruments.

C. TASK: STRAIGHT, CONSTANT AIRSPEED DESCENTS (AMEL and AMES)

REFERENCES: AC 60-14, AC 61-21, AC 61-27; FAA-S-8081-1.

Objective. To determine that the applicant:

1. Exhibits instructional knowledge of the elements of straight, constant airspeed descents, solely by reference to instruments, by describing -

   (a) instrument cross-check, instrument interpretation, and aircraft control.
   (b) instruments used for pitch, bank, and power control during entry, during the descent, and during level-off, and how those instruments are used to maintain descent heading and airspeed.
   (c) trim technique.

2. Exhibits instructional knowledge of common errors related to straight, constant airspeed descents, solely by reference to instruments, by describing -

   (a) "fixation," "omission," and "emphasis" errors during instrument cross-check.
   (b) improper instrument interpretation.
   (c) improper control applications.
   (d) failure to establish proper pitch, bank, or power adjustments during heading and airspeed corrections.
   (e) improper entry or level-off technique.
   (f) faulty trim technique.

3. Demonstrates and simultaneously explains a straight, constant airspeed descent, solely by reference to instruments, from an instructional standpoint.
4. Analyzes and corrects simulated common errors related to straight, constant airspeed descents, solely by reference to instruments.

D. TASK: TURNS TO HEADINGS (AMEL and AMES)

REFERENCES: AC 60-14, AC 61-21, AC 61-27; FAA-S-8081-1.

Objective. To determine that the applicant:

1. Exhibits instructional knowledge of the elements of turns to headings, solely by reference to instruments by describing -

   (a) instrument cross-check, instrument interpretation, and aircraft control.
   (b) instruments used for pitch, bank, and power control during turn entry, during the turn, and during the turn rollout, and how those instruments are used.
   (c) trim technique.

2. Exhibits instructional knowledge of common errors related to turns to headings, solely by reference to instruments, by describing -

   (a) "fixation," "omission," and "emphasis" errors during instrument cross-check.
   (b) improper instrument interpretation.
   (c) improper control applications.
   (d) failure to establish proper pitch, bank, and power adjustments during altitude, bank, and airspeed corrections.
   (e) improper entry or rollout technique.
   (f) faulty trim technique.

3. Demonstrates and simultaneously explains a turn to a heading, solely by reference to instruments, from an instructional standpoint.
4. Analyzes and corrects simulated common errors related to turns to headings, solely by reference to instruments.

E. TASK: RECOVERY FROM UNUSUAL FLIGHT ATTITUDES (AMEL and AMES)

REFERENCES: AC 60-14, AC 61-21, AC 61-27; FAA-S-8081-1.

Objective. To determine that the applicant:

1. Exhibits instructional knowledge of the elements of recovery from unusual flight attitudes by describing -

   (a) conditions and situations that may result in unusual flight attitudes.
   (b) the two basic unusual flight attitudes - nose-high (climbing turn) and nose-low (diving spiral).
   (c) how unusual flight attitudes are recognized.
   (d) control sequence for recovery from a nose-high attitude and the reasons for that sequence.
   (e) control sequence for recovery from a nose-low attitude and the reasons for that sequence.
   (f) reasons why the controls should be coordinated during unusual flight attitude recoveries.

2. Exhibits instructional knowledge of common errors related to recovery from unusual flight attitudes by describing -

   (a) failure to recognize an unusual flight attitude.
   (b) consequences of attempting to recover from an unusual flight attitude by "feel" rather than by instrument indications.
   (c) inappropriate control applications during recovery.
   (d) failure to recognize from instrument indications when the airplane is passing through a level flight attitude.

3. Demonstrates and simultaneously explains recovery from a nose-high and a nose-low unusual flight attitude from an instructional standpoint.
4. Analyzes and corrects simulated common errors related to recovery from unusual flight attitudes.

F. TASK: RADIO AIDS AND RADAR SERVICES (AMEL and AMES)

REFERENCES: AC 60-14, AC 61-21, AC 61-27; FAA-S-8081-1.

Objective. To determine that the applicant:

1. Exhibits instructional knowledge of the elements related to the emergency use of radio aids and radar services by describing -

   (a) situations that would require the use of radio aids or radar assistance.
   (b) available radio aids and radar services.
   (c) how to determine minimum safe altitude.
   (d) procedures to be followed when using a radio aid or radar services.

2. Exhibits instructional knowledge of common errors related to the emergency use of radio aids and radar services by describing -

   (a) hazards of delay in using a radio aid or in obtaining radar services.
   (b) failure to properly control the airplane.
   (c) failure to properly select, tune, or identify a radio facility.
   (d) failure to maintain minimum safe altitude.

3. Demonstrates and simultaneously explains the emergency use of a radio aid or radar services from an instructional standpoint.
4. Analyzes and corrects simulated common errors related to the emergency use of radio aids and radar services.

XII. AREA OF OPERATION: *PERFORMANCE MANEUVERS*

NOTE: The examiner will select TASKS A and B.

A. TASK: STEEP TURNS (AMEL and AMES)

REFERENCES: AC 60-14, AC 61-21; FAA-S-8081-1, FAA-S-8081-2; Pilot's Operating Handbook, FAA-Approved Airplane Flight Manual.

Objective. To determine that the applicant:

1. Exhibits instructional knowledge of the elements of steep turns by describing -

   (a) relationship of bank angle, load factor, and stalling speed.
   (b) overbanking tendency.
   (c) torquo offoct in right and left turns.
   (d) selection of a suitable altitude.
   (e) orientation, division of attention, and planning.
   (f) appropriate power setting and airspeed prior to entry.
   (g) entry and rollout technique.
   (h) coordination of flight controls.
   (i) differential wing drag.
   (j) altitude, bank, and power control during the turn.

2. Exhibits instructional knowledge of common errors related to steep turns by describing -

   (a) improper pitch, bank, and power coordination during entry and rollout.
   (b) uncoordinated use of flight controls.
   (c) inappropriate control applications.
   (d) improper technique in correcting altitude deviations.
   (e) loss of orientation.
   (f) excessive deviation from desired heading during rollout.

3. Demonstrates and simultaneously explains a steep turn from an instructional standpoint.
4. Analyzes and corrects simulated common errors related to steep turns.

B. TASK: EMERGENCY DESCENT (AMEL)

REFERENCES: AC 60-14, AC 61-21; Pilot's Operating Handbook, FAA-Approved Airplane Flight Manual.

Objective. To determine that the applicant:

1. Exhibits instructional knowledge of the elements of an emergency descent by describing -

   (a) situations requiring an emergency descent.
   (b) prescribed configuration of aircraft for emergency descent.
   (c) prescribed speeds in selected configurations as recommended by the manufacturer.
   (d) engine control settings.
   (e) orientation, division of attention, and planning.
   (f) requirement to establish positive load factor during descent.

2. Exhibits instructional knowledge of common errors related to an emergency descent by describing -

   (a) slow response to the emergency.
   (b) failure to establish the prescribed configuration.
   (c) failure to establish and maintain the prescribed speed for the configuration.
   (d) incorrect engine control settings.
   (e) failure to maintain positive load factor in the descent.
   (f) uncoordinated use of controls.

3. Demonstrates and simultaneously explains an emergency descent from an instructional standpoint.
4. Analyzes and corrects simulated common errors related to an emergency descent.

XIII. AREA OF OPERATION: *GROUND REFERENCE MANEUVERS*

NOTE: The examiner will select at least one TASK.

A. TASK: RECTANGULAR COURSE (AMEL and AMES)

REFERENCES: AC 60-14, AC 61-21; FAA-S-8081-1.

Objective. To determine that the applicant:

1. Exhibits instructional knowledge of the elements of a rectangular course by describing -
   - **(a)** how to select a suitable altitude.
   - **(b)** how to select a suitable ground reference with consideration given to emergency landing areas.
   - **(c)** orientation, division of attention, and planning.
   - **(d)** configuration and airspeed prior to entry.
   - **(e)** relationship of a rectangular course to an airport traffic pattern.
   - **(f)** wind drift correction.
   - **(g)** how to maintain desired altitude, airspeed, and distance from ground reference boundaries.
   - **(h)** timing of turn entries and rollouts.
   - **(i)** coordination of flight controls.
2. Exhibits instructional knowledge of common errors related to a rectangular course by describing -
   - **(a)** poor planning, orientation, or division of attention.
   - **(b)** uncoordinated flight control application.
   - **(c)** improper correction for wind drift.
   - **(d)** failure to maintain selected altitude or airspeed.
   - **(e)** selection of a ground reference where there is no suitable emergency landing area within gliding distance.
3. Demonstrates and simultaneously explains a rectangular course from an instructional standpoint.
4. Analyzes and corrects simulated common errors related to a rectangular course.

B. TASK: S-TURNS ACROSS A ROAD (AMEL and AMES)

REFERENCES: AC 60-14, AC 61-21; FAA-S-8081-1.

Objective. To determine that the applicant:

1. Exhibits instructional knowledge of the elements of S-turns across a road by describing -

   (a) how to select a suitable altitude.
   (b) how to select a suitable ground reference line with consideration given to emergency landing areas.
   (c) orientation, division of attention, and planning.
   (d) configuration and airspeed prior to entry.
   (e) entry technique.
   (f) wind drift correction.
   (g) tracking of semicircles of equal radii on either side of the selected ground reference line.
   (h) how to maintain desired altitude and airspeed.
   (i) turn reversal over the ground reference line.
   (j) coordination of flight controls.

2. Exhibits instructional knowledge of common errors related to S-turns across a road by describing -

   (a) faulty entry technique.
   (b) poor planning, orientation, or division of attention.
   (c) uncoordinated flight control application.
   (d) improper correction for wind drift.
   (e) an unsymmetrical ground track.
   (f) failure to maintain selected altitude or airspeed.
   (g) selection of a ground reference line where there is no suitable emergency landing area within gliding distance.

3. Demonstrates and simultaneously explains S-turns across a road from an instructional standpoint.
4. Analyzes and corrects simulated common errors related to S-turns across a road.

C. TASK: TURNS AROUND A POINT (AMEL and AMES)

REFERENCES: AC 60-14, AC 61-21; FAA-S-8081-1.

Objective. To determine that the applicant:

1. Exhibits instructional knowledge of the elements of turns around a point by describing -

   (a) how to select a suitable altitude.
   (b) how to select a suitable ground reference point with consideration given to emergency landing areas.
   (c) orientation, division of attention, and planning.
   (d) configuration and airspeed prior to entry.
   (e) entry technique.
   (f) wind drift correction.
   (g) how to maintain desired altitude, airspeed, and distance from reference point.
   (h) coordination of flight controls.

2. Exhibits instructional knowledge of common errors related to turns around a point by describing -

   (a) faulty entry technique.
   (b) poor planning, orientation, or division of attention.
   (c) uncoordinated flight control application.
   (d) improper correction for wind drift.
   (e) failure to maintain selected altitude or airspeed.
   (f) selection of a ground reference point where there is no suitable emergency landing area within gliding distance.

3. Demonstrates and simultaneously explains turns around a point from an instructional standpoint.
4. Analyzes and corrects simulated common errors related to turns around a point.

XIV. AREA OF OPERATION: *EMERGENCY OPERATIONS*

NOTE: The examiner will select TASKS A, B, C, D, E, F, and G.

A. TASK: SYSTEMS AND EQUIPMENT MALFUNCTIONS (AMEL and AMES)

REFERENCES: AC 61-21; FAA-S-8081-1, FAA-S-8081-2; Pilot's Operating Handbook, FAA-Approved Airplane Flight Manual.

NOTE: The examiner will not simulate a system or equipment malfunction in a manner that may jeopardize safe flight or result in possible damage to the airplane.

Objective. To determine that the applicant exhibits instructional knowledge of the elements related to systems and equipment malfunctions, appropriate to the airplane used for the practical test, by describing recommended pilot action for:

1. Smoke, fire, or both, during ground or flight operations.
2. Rough running engine, partial power loss, or sudden engine stoppage.
3. Propeller malfunction.
4. Loss of engine oil pressure.
5. Fuel starvation.
6. Engine overheat.
7. Hydraulic system malfunction.
8. Electrical system malfunction.
9. Carburetor or induction icing.
10. Door or window opening in flight.
11. Inoperative or "runaway" trim.
12. Landing gear or flap malfunction.
13. Pressurization malfunction.
14. Any other system or equipment malfunction.

B. TASK: MANEUVERING WITH ONE ENGINE INOPERATIVE (AMEL and AMES)

REFERENCES: AC 60-14, AC 61-21; FAA-S-8081-1, FAA-S-8081-2; Pilot's Operating Handbook, FAA-Approved Airplane Flight Manual.

NOTE: The feathering of one propeller shall be demonstrated in any multiengine airplane equipped with propellers which can be safely feathered and unfeathered in flight. Feathering for pilot flight test purposes should be performed only under such conditions and at such altitudes (no lower than 3,000 feet/1,000 meters above the surface) and positions where safe landings on established airports can be readily accomplished, in the event difficulty is encountered in unfeathering. At altitudes lower than 3,000 feet above the surface, simulated engine failure will be performed by throttling the engine and then establishing zero thrust.

In the event a propeller cannot be unfeathered during the practical test, it should be treated as an emergency.

Objective. To determine that the applicant:

1. Exhibits instructional knowledge of the elements related to maneuvering with one engine inoperative by describing -
   - **(a)** flight characteristics and controllability associated with maneuvering with one engine inoperative.
   - **(b)** use of prescribed emergency checklist to verify accomplishment of procedures for securing inoperative engine.
   - **(c)** proper adjustment of engine controls, reduction of drag, and identification and verification of the inoperative engine.
   - **(d)** how to establish and maintain the best engine inoperative airspeed.
   - **(e)** proper trim technique.
   - **(f)** how to establish and maintain a bank, as required, for best performance.

(g) appropriate methods to be used for determining the reason for the malfunction.
(h) importance of establishing a heading toward the nearest suitable airport or seaplane base.
(i) importance of monitoring and adjusting the operating engine.
(j) performance of straight-and-level flight, turns, descents, and climbs, if the airplane is capable of those maneuvers under existing conditions.

2. Exhibits instructional knowledge of common errors related to maneuvering with one engine inoperative by describing -

   (a) failure to follow prescribed emergency checklist.
   (b) failure to recognize an inoperative engine.
   (c) hazards of improperly identifying and verifying the inoperative engine.
   (d) failure to properly adjust engine controls and reduce drag.
   (e) failure to establish and maintain the best engine inoperative airspeed.
   (f) improper trim technique.
   (g) failure to establish and maintain proper bank for best performance.
   (h) failure to maintain positive control while maneuvering.
   (i) hazards of attempting flight contrary to the airplane's operating limitations.

3. Demonstrates and simultaneously explains maneuvering with one engine inoperative from an instructional standpoint.
4. Analyzes and corrects simulated common errors related to maneuvering with one engine inoperative.

C. TASK: ENGINE INOPERATIVE LOSS OF DIRECTIONAL CONTROL DEMONSTRATION (AMEL and AMES)

REFERENCES: AC 60-14, AC 61-21; FAA-S-8081-1, FAA-S-8081-2; Pilot's Operating Handbook, FAA-Approved Airplane Flight Manual.

NOTE: FAR Part 1 defines V_{MC} as minimum control speed with the critical engine inoperative. There is a density altitude above which the stalling speed is higher than the engine inoperative minimum control speed. When this density altitude exists close to the ground because of high elevations, high temperatures, or both, an effective flight demonstration of loss of directional control may be hazardous and should not be attempted. If it is determined prior to flight that the stall speed is above or equal to V_{MC}, this flight demonstration is impracticable. In this case, the significance of the engine inoperative minimum control speed should be emphasized through oral questioning, including the results of attempting engine inoperative flight below that speed, the recognition of loss of directional control, and proper recovery techniques.

Recovery should be made by simultaneously reducing the power on the operating engine and reducing the angle of attack as necessary to regain directional control and airspeed. Recoveries should not be attempted by increasing power on the simulated failed engine.

Performing this maneuver by increasing pitch attitude to a high angle with both engines operating and then reducing power on the critical engine should be avoided. This technique is hazardous and may result in loss of aircraft control

Objective. To determine that the applicant:

1. Exhibits instructional knowledge of the elements related to engine inoperative loss of directional control by describing -

   (a) causes of loss of directional control at airspeeds less than V_{MC}, the factors affecting V_{MC}, and the safe recovery procedures.
   (b) establishment of airplane configuration, adjustment of power controls, and trim prior to the demonstration.

(c) establishment of engine inoperative pitch attitude and airspeed.
(d) establishment of a bank attitude as required for best performance.
(e) entry technique to demonstrate loss of directional control.
(f) indications that enable a pilot to recognize loss of directional control.
(g) proper recovery technique.

2. Exhibits instructional knowledge of common errors related to engine inoperative loss of directional control by describing -

(a) inadequate knowledge of the causes of loss of directional control at airspeeds less than V_{MC}, factors affecting V_{MC}, and safe recovery procedures.
(b) improper entry procedures, including pitch attitude, bank attitude, and airspeed.
(c) failure to recognize imminent loss of directional control.
(d) failure to use proper recovery technique.
(e) rough and/or uncoordinated control technique.

3. Demonstrates and simultaneously explains engine inoperative loss of directional control from an instructional standpoint.
4. Analyzes and corrects simulated common errors related to engine inoperative loss of directional control.

D. TASK: DEMONSTRATING THE EFFECTS OF VARIOUS AIRSPEEDS AND CONFIGURATIONS DURING ENGINE INOPERATIVE PERFORMANCE (AMEL and AMES)

REFERENCES: AC 60-14, AC 61-21; FAA-S-8081-1, FAA-S-8081-2; Pilot's Operating Handbook, FAA-Approved Airplane Flight Manual.

Objective. To determine that the applicant:

1. Exhibits instructional knowledge of the elements related to the effects of various airspeeds and configurations during engine inoperative performance by describing -

   (a) selection of proper altitude for the demonstration.
   (b) proper entry procedure to include pitch attitude, bank attitude, and airspeed.
   (c) effects on performance of airspeed changes at, above, and below V_{YSE}.
   (d) effects on performance of various configurations -

      (1) extension of landing gear.
      (2) extension of wing flaps.
      (3) extension of both landing gear and wing flaps.
      (4) windmilling of propeller on inoperative engine.

   (e) airspeed control throughout the demonstration.
   (f) proper control technique and procedures throughout the demonstration.

2. Exhibits instructional knowledge of common errors related to the effects of various airspeeds and configurations during engine inoperative performance by describing -

   (a) inadequate knowledge of the effects of airspeeds above or below V_{YSE} and of various configurations on performance.
   (b) improper entry procedures, including pitch attitude, bank attitude, and airspeed.

(c) improper airspeed control throughout the demonstration.
(d) rough and/or uncoordinated control technique.
(e) improper procedures during resumption of cruise flight.

3. Demonstrates and simultaneously explains the effects of various airspeeds and configurations during engine inoperative performance from an instructional standpoint.
4. Analyzes and corrects simulated common errors related to the effects of various airspeeds and configurations during engine inoperative performance.

E. TASK: ENGINE FAILURE DURING TAKEOFF BEFORE V_{MC} (AMEL and AMES)

NOTE: Engine failure will not be simulated at a speed greater than 50 percent V_{MC}.

REFERENCES: AC 60-14, AC 61-21; FAA-S-8081-1, FAA-S-8081-2; Pilot's Operating Handbook, FAA-Approved Airplane Flight Manual.

Objective. To determine that the applicant:

1. Exhibits instructional knowledge of the elements related to engine failure during takeoff before V_{MC} by describing -

(a) use of prescribed emergency procedure.
(b) prompt closing of throttles.
(c) how to maintain directional control.
(d) proper use of brakes (landplane).

2. Exhibits instructional knowledge of common errors related to engine failure during takeoff before V_{MC} by describing -

(a) failure to follow prescribed emergency procedure.
(b) failure to promptly recognize engine failure.
(c) failure to promptly close throttles following engine failure.
(d) faulty directional control and use of brakes.

3. Demonstrates and simultaneously explains a simulated engine failure during takeoff before V_{MC} from an instructional standpoint.

4. Analyzes and corrects simulated common errors related to engine failure during takeoff before V_{MC}.

F. TASK: ENGINE FAILURE AFTER LIFT-OFF (AMEL and AMES)

REFERENCES: AC 60-14, AC 61-21; FAA-S-8081-1, FAA-S-8081-2; Pilot's Operating Handbook, FAA-Approved Airplane Flight Manual.

Objective. To determine that the applicant:

1. Exhibits instructional knowledge of the elements related to engine failure after lift-off by describing -

   (a) use of prescribed emergency checklist to verify accomplishment of procedures for securing the inoperative engine.
   (b) proper adjustment of engine controls, reduction of drag, and identification and verification of the inoperative engine.
   (c) how to establish and maintain a pitch attitude that will result in the best engine inoperative airspeed, considering the height of obstructions.
   (d) how to establish and maintain a bank as required for best performance.
   (e) how to maintain directional control.
   (f) methods to be used for determining reason for malfunction.
   (g) monitoring and proper use of the operating engine.
   (h) an emergency approach and landing, if a climb or level flight is not within the airplane's performance capability.
   (i) positive airplane control.
   (j) how to obtain assistance from the appropriate facility.

2. Exhibits instructional knowledge of common errors related to engine failure after lift-off by describing -

   (a) failure to follow prescribed emergency checklist.
   (b) failure to properly identify and verify the inoperative engine.
   (c) failure to properly adjust engine controls and reduce drag.
   (d) failure to maintain directional control.
   (e) failure to establish and maintain a pitch attitude that will result in best engine inoperative airspeed, considering the height of obstructions.
   (f) failure to establish and maintain proper bank for best performance.

3. Demonstrates and simultaneously explains a simulated engine failure after lift-off from an instructional standpoint.
4. Analyzes and corrects simulated common errors related to engine failure after lift-off.

G. TASK: APPROACH AND LANDING WITH AN INOPERATIVE ENGINE (AMEL and AMES)

REFERENCES: AC 60-14, AC 61-21; FAA-S-8081-1, FAA-S-8081-2; Pilot's Operating Handbook, FAA-Approved Airplane Flight Manual.

Objective. To determine that the applicant:

1. Exhibits instructional knowledge of the elements related to an approach and landing with an inoperative engine by describing -

   (a) use of the prescribed emergency checklist to verify accomplishment of procedures for securing the inoperative engine.
   (b) proper adjustment of engine controls, reduction of drag, and identification and verification of the inoperative engine.
   (c) how to establish and maintain best engine inoperative airspeed.
   (d) trim technique.
   (e) how to establish and maintain a bank as required for best performance.

(f) the monitoring and adjusting of the operating engine.
(g) proper approach to selected touchdown area, at the recommended airspeed.
(h) proper application of flight controls.
(i) how to maintain a precise ground track.
(j) wind shear and turbulence.
(k) proper timing, judgment, and control technique during roundout and touchdown.
(l) directional control after touchdown.
(m) use of brakes (landplane).

2. Exhibits instructional knowledge of common errors related to an approach and landing with an inoperative engine by describing -

(a) failure to follow prescribed emergency checklist.
(b) failure to properly identify and verify the inoperative engine.
(c) failure to properly adjust engine controls and reduce drag.
(d) failure to establish and maintain best engine inoperative airspeed.
(e) improper trim technique.
(f) failure to establish proper approach and landing configuration at appropriate time and in proper sequence.
(g) failure to use proper technique for wind shear or turbulence.
(h) inappropriate removal of hand from throttles.
(i) faulty technique during roundout and touchdown.
(j) improper directional control after touchdown.
(k) improper use of brakes (landplane).

3. Demonstrates and simultaneously explains an approach and landing with a simulated inoperative engine from an instructional standpoint.

4. Analyzes and corrects simulated common errors related to an approach and landing with an inoperative engine.

H. TASK: EMERGENCY EQUIPMENT AND SURVIVAL GEAR (AMEL and AMES)

REFERENCES: AC 61-21; FAA-S-8081-1, FAA-S-8081-2; Pilot's Operating Handbook, FAA-Approved Airplane Flight Manual.

Objective. To determine that the applicant exhibits instructional knowledge of the elements related to emergency equipment and survival gear appropriate to the airplane flown by describing:

1. Locations in the airplane.
2. Purpose.
3. Method of operation or use.
4. Servicing.
5. Storage.
6. Equipment and gear appropriate for operation in various climates, over various types of terrain, and over water.

XV. AREA OF OPERATION: *APPROACHES AND LANDINGS*

NOTE: The examiner will select at least one TASK.

A. TASK: NORMAL AND CROSSWIND APPROACH AND LANDING (AMEL and AMES)

REFERENCES: AC 60-14, AC 61-21; FAA-S-8081-1, FAA-S-8081-2; Pilot's Operating Handbook, FAA-Approved Airplane Flight Manual, Seaplane Manual.

Objective. To determine that the applicant:

1. Exhibits instructional knowledge of the elements of a normal and a crosswind approach and landing by describing -

   (a) how to determine landing performance and limitations.
   (b) configuration, power, and trim.
   (c) obstructions and other hazards which should be considered.
   (d) a stabilized approach at the recommended airspeed to the selected touchdown area.
   (e) coordination of flight controls.
   (f) a precise ground track.
   (g) wind shear and wake turbulence.
   (h) most suitable crosswind technique.
   (i) timing, judgment, and control technique during roundout and touchdown.
   (j) directional control after touchdown.
   (k) use of brakes (landplane).
   (l) use of checklist.

2. Exhibits instructional knowledge of common errors related to a normal and a crosswind approach and landing by describing -

   (a) improper use of landing performance data and limitations.
   (b) failure to establish approach and landing configuration at appropriate time or in proper sequence.

(c) failure to establish and maintain a stabilized approach.
(d) inappropriate removal of hand from throttles.
(e) improper technique during roundout and touchdown.
(f) poor directional control after touchdown.
(g) improper use of brakes (landplane).

3. Demonstrates and simultaneously explains a normal or a crosswind approach and landing from an instructional standpoint.
4. Analyzes and corrects simulated common errors related to a normal or crosswind approach and landing.

B. TASK: GO-AROUND (AMEL and AMES)

REFERENCES: AC 60-14, AC 61-21; FAA-S-8081-1, FAA-S-8081-2; Pilot's Operating Handbook, FAA-Approved Airplane Flight Manual.

Objective. To determine that the applicant:

1. Exhibits instructional knowledge of the elements of a go-around by describing -

(a) situations where a go-around is necessary.
(b) importance of making a prompt decision.
(c) importance of applying takeoff power immediately after the go-around decision is made.
(d) importance of establishing proper pitch attitude.
(e) wing flaps retraction.
(f) use of trim.
(g) landing gear retraction.
(h) proper climb speed.
(i) proper track and obstruction clearance.
(j) use of checklist.

2. Exhibits instructional knowledge of common errors related to a go-around by describing -

(a) failure to recognize a situation where a go-around is necessary.
(b) hazards of delaying a decision to go around.
(c) improper power application.

(d) failure to control pitch attitude.
(e) failure to compensate for torque effect.
(f) improper trim technique.
(g) failure to maintain recommended airspeeds.
(h) improper wing flaps or landing gear retraction procedure.
(i) failure to maintain proper track during climb-out.
(j) failure to remain well clear of obstructions and other traffic.

3. Demonstrates and simultaneously explains a go-around from an instructional standpoint.
4. Analyzes and corrects simulated common errors related to a go-around.

C. TASK: SHORT-FIELD APPROACH AND LANDING (AMEL)

REFERENCES: AC 60-14, AC 61-21; FAA-S-8081-1, FAA-S-8081-2; Pilot's Operating Handbook, FAA-Approved Airplane Flight Manual.

Objective. To determine that the applicant:

1. Exhibits instructional knowledge of the elements of a short-field approach and landing by describing -

   (a) how to determine landing performance and limitations.
   (b) configuration and trim.
   (c) proper use of pitch and power to maintain desired approach angle.
   (d) barriers and other hazards which should be considered.
   (e) effect of wind.
   (f) selection of touchdown and go-around points.
   (g) a stabilized approach at the recommended airspeed to the selected touchdown point.
   (h) coordination of flight controls.
   (i) a precise ground track.
   (j) timing, judgment, and control technique during roundout and touchdown.
   (k) directional control after touchdown.
   (l) use of brakes.
   (m) use of checklist.

2. Exhibits instructional knowledge of common errors related to a short-field approach and landing by describing -

   (a) improper use of landing performance data and limitations.
   (b) failure to establish approach and landing configuration at appropriate time or in proper sequence.
   (c) failure to establish and maintain a stabilized approach.
   (d) improper technique in use of power, wing flaps, and trim.
   (e) inappropriate removal of hand from throttles.
   (f) improper technique during roundout and touchdown.
   (g) poor directional control after touchdown.
   (h) improper use of brakes.

3. Demonstrates and simultaneously explains a short-field approach and landing from an instructional standpoint.
4. Analyzes and corrects simulated common errors related to a short-field approach and landing.

D. TASK: GLASSY-WATER APPROACH AND LANDING (AMES)

REFERENCES: AC 60-14, AC 61-21; FAA-S-8081-1, FAA-S-8081-2; Pilot's Operating Handbook, FAA-Approved Airplane Flight Manual, Seaplane Manual.

Objective. To determine that the applicant:

1. Exhibits instructional knowledge of the elements of a glassy-water approach and landing by describing -

   (a) how to determine landing performance and limitations.
   (b) configuration and trim.
   (c) deceptive characteristics of glassy water.
   (d) selection of a suitable landing area and go-around point.
   (e) terrain and obstructions which should be considered.

(f) detection of hazards in the landing area such as shallow water, obstructions, or floating debris.
(g) coordination of flight controls.
(h) a precise ground track.
(i) a power setting and pitch attitude that will result in the recommended airspeed and rate of descent throughout the final approach to touchdown.
(j) how to maintain positive after-landing control.
(k) use of checklist.

2. Exhibits instructional knowledge of common errors related to a glassy-water approach and landing by describing -

(a) improper use of landing performance data and limitations.
(b) failure to establish approach and landing configuration at appropriate time and in proper sequence.
(c) failure to establish and maintain a stabilized approach at the recommended airspeed and rate of descent.
(d) improper technique in use of power, wing flaps, or trim.
(e) inappropriate removal of hand from throttles.
(f) failure to touch down with power in the proper stabilized attitude.
(g) failure to maintain positive after-landing control.

3. Demonstrates and simultaneously explains a glassy-water approach and landing from an instructional standpoint.
4. Analyzes and corrects simulated common errors related to a glassy-water approach and landing.

E. TASK: ROUGH-WATER APPROACH AND LANDING (AMES)

REFERENCES: AC 60-14, AC 61-21; FAA-S-8081-1, FAA-S-8081-2; Pilot's Operating Handbook, FAA-Approved Airplane Flight Manual, Seaplane Manual.

Objective. To determine that the applicant:

1. Exhibits instructional knowledge of the elements of a rough-water approach and landing by describing -

   (a) determination of landing performance and limitations.
   (b) review of wind conditions.
   (c) how landing area characteristics can be evaluated.
   (d) selection of a suitable landing area and go-around point.
   (e) terrain and obstructions which should be considered.
   (f) detection of hazards in the landing area such as shallow water, obstructions, or floating debris.
   (g) configuration and trim.
   (h) coordination of flight controls.
   (i) a precise ground track.
   (j) a stabilized approach at the recommended airspeed to the selected touchdown area.
   (k) timing, judgment, and control technique during roundout and touchdown.
   (l) maintenance of positive after-landing control.
   (m) use of checklist.

2. Exhibits instructional knowledge of common errors related to a rough-water approach and landing by describing -

   (a) improper use of landing performance data and limitations.
   (b) failure to establish approach and landing configuration at appropriate time and in proper sequence.
   (c) failure to establish and maintain a stabilized approach.
   (d) improper technique in use of power, wing flaps, or trim.

(e) inappropriate removal of hand from throttles.
(f) improper technique during roundout and touchdown.
(g) failure to maintain positive after-landing control.

3. Demonstrates and simultaneously explains a rough-water approach and landing from an instructional standpoint.
4. Analyzes and corrects simulated common errors related a rough-water approach and landing.

F. TASK: CONFINED-AREA APPROACH AND LANDING (AMES)

REFERENCES: AC 60-14, AC 61-21; FAA-S-8081-1, FAA-S-8081-2; Pilot's Operating Handbook, FAA-Approved Airplane Flight Manual, Seaplane Manual.

Objective. To determine that the applicant:

1. Exhibits instructional knowledge of the elements of a confined-area approach and landing by describing -

   (a) how to determine landing performance and limitations.
   (b) approaches and landings on various types of water areas.
   (c) effect of wind and water condition.
   (d) selection of a suitable landing area and go-around point.
   (e) terrain and obstructions which should be considered.
   (f) detection of hazards in the landing area such as shallow water, obstructions, or floating debris.
   (g) configuration and trim.
   (h) a stabilized approach at the recommended airspeed to the selected touchdown area.
   (i) coordination of flight controls.
   (j) a precise ground track.
   (k) timing, judgment, and control technique during roundout and touchdown.
   (l) touchdown in the proper pitch attitude at the minimum safe airspeed.
   (m) how to maintain positive after-landing control.
   (n) use of checklist.

2. Exhibits instructional knowledge of common errors related to a confined-area approach and landing by describing -

   (a) improper use of landing performance data and limitations.
   (b) failure to establish approach and landing configuration at appropriate time and in proper sequence.
   (c) failure to establish and maintain a stabilized approach.
   (d) improper technique in the use of power, wing flaps, and trim.
   (e) inappropriate removal of hand from throttles.
   (f) improper technique during roundout and touchdown.
   (g) failure to maintain positive after-landing control.

3. Demonstrates and simultaneously explains a confined-area approach and landing from an instructional standpoint.
4. Analyzes and corrects simulated common errors related to a confined-area approach and landing.

XVI. AREA OF OPERATION: *AFTER-LANDING PROCEDURES*

NOTE: The examiner will select at least TASK E.

A. TASK: ANCHORING (AMES)

REFERENCES: AC 60-14, AC 61-21; FAA-S-8081-1, FAA-S-8081-2; Seaplane Manual.

Objective. To determine that the applicant:

1. Exhibits instructional knowledge of the elements of anchoring by describing -

   (a) how to select a suitable area for anchoring.
   (b) recommended procedure for anchoring in a lake, river, or tidal area.
   (c) number of anchors and lines to be used to ensure seaplane security in various conditions.
   (d) hazards to be avoided.

2. Exhibits instructional knowledge of common errors related to anchoring by describing -

   (a) hazards resulting from failure to follow recommended procedures.
   (b) consequences of poor planning, improper technique, or poor judgment when anchoring.
   (c) consequences of failure to use anchor lines of adequate length and strength to ensure seaplane security.

3. Demonstrates and simultaneously explains anchoring from an instructional standpoint.
4. Analyzes and corrects simulated common errors related to anchoring.

B. TASK: DOCKING AND MOORING (AMES)

REFERENCES: AC 60-14, AC 61-21; FAA-S-8081-1, FAA-S-8081-2; Seaplane Manual.

Objective. To determine that the applicant:

1. Exhibits instructional knowledge of the elements of docking and mooring by describing -

   (a) recommended procedures for docking.
   (b) recommended procedures for mooring.
   (c) hazards to be considered when docking and mooring.
   (d) procedures to be followed to ensure seaplane security.
   (e) requirement for mooring lights.

2. Exhibits instructional knowledge of common errors related to docking and mooring by describing -

   (a) hazards resulting from failure to follow recommended procedures.
   (b) consequences of poor planning, improper technique, or poor judgment when docking and mooring.
   (c) consequences of failure to take appropriate precautions to avoid hazards or to ensure that seaplane is secure.

3. Demonstrates and simultaneously explains docking and mooring from an instructional standpoint.
4. Analyzes and corrects simulated common errors related to docking and mooring.

C. TASK: BEACHING (AMES)

REFERENCES: AC 60-14, AC 61-21; FAA-S-8081-1, FAA-S-8081-2; Seaplane Manual.

Objective. To determine that the applicant:

1. Exhibits instructional knowledge of the elements of beaching by describing -

   (a) recommended procedures for beaching.
   (b) factors to be considered such as beach selection, water depth, current, tide, and wind.
   (c) procedures to be followed to ensure seaplane security.
   (d) hazards to be avoided.

2. Exhibits instructional knowledge of common errors related to beaching by describing -

   (a) hazards resulting from failure to follow recommended procedures.
   (b) consequences of poor beach selection, poor planning, improper technique, or faulty judgment when beaching.
   (c) consequences of failure to take appropriate precautions to avoid hazards or to ensure that seaplane is secure.

3. Demonstrates and simultaneously explains beaching from an instructional standpoint.
4. Analyzes and corrects simulated common errors related to beaching.

D. TASK: **RAMPING** (AMES)

REFERENCES: AC 60-14, AC 61-21; FAA-S-8081-1, FAA-S-8081-2; Seaplane Manual.

Objective. To determine that the applicant:

1. Exhibits instructional knowledge of the elements of ramping by describing -

   (a) factors such as type of ramp surface, wind, current, and direction and control of approach speed.
   (b) recommended procedures for ramping.
   (c) hazards to be avoided.

2. Exhibits instructional knowledge of common errors related to ramping by describing -

   (a) hazards resulting from failure to follow recommended procedures.
   (b) consequences of poor planning, improper technique, or faulty judgment when ramping.
   (c) consequences of failure to take appropriate precautions to avoid hazards or to ensure that the seaplane is secure.

3. Demonstrates and simultaneously explains ramping from an instructional standpoint.
4. Analyzes and corrects simulated common errors related to ramping.

E. TASK: **POSTFLIGHT PROCEDURES** (AMEL and AMES)

REFERENCES: AC 60-14, AC 61-21; FAA-S-8081-1, FAA-S-8081-2; Pilot's Operating Handbook, FAA-Approved Airplane Flight Manual, Seaplane Manual.

Objective. To determine that the applicant:

1. Exhibits instructional knowledge of the elements of postflight procedures by describing -

   (a) parking technique and procedure (landplane).
   (b) engine shutdown and securing cockpit.

(c) deplaning passengers.
(d) securing airplane.
(e) postflight inspection.
(f) refueling.

2. Exhibits instructional knowledge of common errors related to postflight procedures by describing -

(a) hazards resulting from failure to follow recommended procedures.
(b) poor planning, improper technique, or faulty judgment in performance of postflight procedures.

3. Demonstrates and simultaneously explains postflight procedures from an instructional standpoint.
4. Analyzes and corrects simulated common errors related to postflight procedures.

☆ U.S. G.P.O. 1991 522-028/40074